Emplacing Food

Series Editors

Emplacing Food critically investigates the intricate relationships between food, space, and society, with particular emphasis on urban and local scales. This series publishes interdisciplinary works that contribute to a nuanced understanding of food systems as they operate within diverse socio-cultural and environmental contexts.

Seeking to foster a comprehensive analysis of the interconnections between food practices and spatial dynamics, the series emphasizes the necessity for integrated food policies that address pressing issues such as food security, access, and justice. Each volume provides actionable insights that informs the creation of new food policies. Topics include the food-health nexus, environmental impacts of food systems, technological innovations in food chains, and cultural identities associated with food.

The series welcomes contributions from a wide range of disciplines within the social sciences and humanities, enabling a holistic exploration of food systems, encompassing economic, social, cultural, and environmental dimensions.

Paola Minoia
Editor

Foodways as Sites of Knowledge and Transformation

Researching Meanings, Relationalities and Space

palgrave
macmillan

Editor
Paola Minoia
Department of Cultures, Politics and Society
University of Turin
Turin, Italy

Emplacing Food
ISBN 978-3-032-28125-8 ISBN 978-3-032-28126-5 (eBook)
https://doi.org/10.1007/978-3-032-28126-5

This work was supported by the Italian Ministry of University and Research under Grant number P20223SFMN.

This Palgrave Macmillan imprint is published by the registered company Springer Nature Switzerland AG
The registered company address is: Gewerbestrasse 11, 6330 Cham, Switzerland

Contents

Notes on Contributors

Beatrice Ferlaino is a research fellow at the University of Turin. Her work examines food practices across the entire chain, from agricultural production and fishing to consumption, including distribution, prices, and markets. She investigates power dynamics, governmentality, and the paradigms shaping political and individual choices, as well as processes of legitimation. Her research addresses socio-environmental change and its implications for food access and inequality. She adopts an interdisciplinary approach that bridges the historical sociology of political processes with political and social geography across diverse contemporary contexts.

Paola Minoia is an Associate Professor of Geography at the Department of Cultures, Politics and Society of the University of Turin and an Adjunct Professor at the University of Helsinki. Her research focuses on the political ecology of water and food, postdevelopment and the coloniality of international cooperation, the recognition of Indigenous knowledges, environmental justice, and tourism-driven gentrification in cities, with an emphasis on relational methods in geographical research, and numerous publications in these fields. She is the Principal Investigator of the research project *Post-development Geographies of Local Food Systems: Community-based Networks Addressing Food Insecurity*, from which this book originated.

Ginevra Montefusco is a Ph.D. student in Geography at the University of Padua and Ca' Foscari of Venice. Her thesis focuses on artisanal fishing as a relational practice connecting urban and marine ecologies in Tangier and Naples. Her research interests include food geographies (foodways and place-making), urban geographies (Mediterranean, Arab and Southern urbanism), and geographies of the sea (ontological and more-than-human approaches). Her research practice intersects visual arts and transformative methods, including photography, ethnographic video-making, and collective counter-mapping.

Tamara Taher is a post-doctoral researcher at the Department of Cultures, Politics and Society at the University of Turin. Her research focuses on questions of temporality, epistemology, food geographies, and politics. She adopts a multi-disciplinary approach, moving between geography, social and political theory, IR, and anthropology. She has consistently worked on food practices and knowledges in Palestine and Jordan. Her research interests also revolve around structural racism and anti-racism in European contexts, memory, history, vernacular and embodied knowledges and radical pedagogies.

CHAPTER 1

Introduction

Paola Minoia

Abstract This introductory chapter frames the volume within critical food studies and political ecology by conceptualizing foodways as relational practices through which human and more-than-human lives are mutually shaped. Based on empirically grounded, participatory, and community-based research, the contributions explore fishing, agriculture, food preparation, and meal sharing as sites of knowledge production, ethical engagement, and social reproduction. Food systems are approached as foodscapes: spatial and political formations shaped by unequal access to land, water, and labour within broader regimes of power and global market dynamics, within which communities enact situated forms of agency. Challenging dominant technoscientific and development-oriented narratives, the volume advances a relational and postdevelopment perspective that foregrounds local knowledges, multispecies subjectivities, and everyday practices. Drawing on feminist, postcolonial, and radical food geography approaches, it examines how communities navigate ecological change, urbanization, and extractivism.

P. Minoia (✉)
Department of Cultures, Politics and Society, University of Turin, Turin 10153, Italy
e-mail: paola.minoia@unito.it

P. Minoia (ed.), *Foodways As Sites of Knowledge and Transformation*,
Emplacing Food, https://doi.org/10.1007/978-3-032-28126-5_1

Through case studies in Morocco and Jordan, with a focus on artisanal fishing and urban food initiatives, the chapters highlight forms of cooperation, learning, and collective negotiation. Rather than proposing universal solutions, the book offers a situated and reflexive approach to research grounded in long-term engagement, inviting readers to rethink food systems as sites of justice and more liveable multispecies futures.

Keywords Foodways · Multispecies relations · Community agency · Relational practices · Postdevelopment

The volume brings together empirically grounded contributions that focus on foodways, understood as practices of fishing, agriculture, food preparation, and meal sharing through which human and nonhuman lives are mutually shaped. Drawing on sensory, experimental, participatory, and community-based research, the chapters attend to the onto-epistemological and ethical dimensions of working-with, caring-for, and relating-to more-than-human worlds. Rather than treating food as a mere object of consumption or production, the contributors conceptualize foodways as relational microcosms in which affective, cultural, political, and ecological exchanges take place.

The spatial dimension is central to this analysis and encompasses the ecologies and infrastructures that constitute the substratum of food production. Within this spatial consideration, foodscapes, as the landscapes through which food becomes visible and actionable, are in turn both material and political, as they are forged through historically situated relations of power and uneven access to resources.

A central concern running throughout the book is how the concrete engagements of local communities with fish, soils, seawaters, crops, and the material conditions of food production, shape modes of perception, thought, and collaboration. From this perspective, the volume addresses a set of interrelated questions that guide the contributions.

What forms of attention, responsibility, and mutuality emerge through everyday foodways? How do specific nonhuman subjectivities, such as plants attuned to seasonal rhythms, fish shaped by patterns of mobility, soils bearing the traces of care, drought, and depletion, and ecosystems undergoing rapid ecological change, together with human communities increasingly constrained by processes of urbanization and extractivism, shape agricultural and fishing practices, livelihoods, and food systems? And how can we approach the field in ways attentive to ongoing ecological change, local forms of agency, and political imaginaries, while rethinking our responsibilities as researchers and citizens?

These questions call for a fundamental shift in research perspective within the current polycrisis. They point to the need to reconsider not only what kinds of knowledge are produced, but also the assumptions, hierarchies, and colonial legacies through which research engages with communities in the global majority world. It is from this critical standpoint that the origins of this book must be understood.

In fact, the volume originates from a research project developed in response to a funding call aimed at delivering novel technoscientific solutions to food scarcity through applied research. That call was aligned with dominant development narratives in which systemic change is imagined as a linear transition towards technological modernization, driven by capital investment and facilitated by the entry of external corporations into local territories. When applied to the MENA context, where the authors worked through long-standing engagements with local community groups, this understanding of technoscientific intervention becomes particularly visible in the language and practices of aid and international cooperation, imbued with forms of coloniality that marginalize local knowledges in the name of presumed European epistemic superiority.

What technoscientific solutions systematically fail to address is that food is deeply entangled in political-economic relations, global trade regimes, and national decisions concerning productive specialization. These processes often operate to the detriment of small-scale producers and niche crops, which are crucial for the preservation of biological and cultural diversity, yet are progressively marginalized and squeezed out of dominant food systems. Additional pressures are exerted by national political elites who frame development primarily in terms of urban and infrastructural expansion, resulting in rapid land consumption, the dismantling of traditional subsistence systems, and increasingly restricted access to primary resources such as land and water.

Food systems thus actively participate in the political construction of power relations that become materially visible in what political ecologists have described as landscapes of power (Mitchell, 2002; Robbins, 2012). A striking illustration can be found, for instance, in the vast plains of Morocco now covered by greenhouse agriculture oriented towards export markets. These are not food productions intended for local consumption, but highly spatialized regimes of production that take shape where international capital is able to negotiate access to land, water, and labour at low cost: what Patel and Moore (2017) conceptualize as the making of "cheap natures" and "cheap lives" within capitalist world-ecologies.

Against this backdrop, this book contributes to a growing body of scholarship at the intersection of political ecology and critical food studies by foregrounding foodways as privileged arenas for exploring multispecies relations, situated ways of knowing, and struggles for socio-ecological justice. Our approach deliberately diverges from those technoscientific framings. We responded to the call not by proposing technical solutions or predefined models of food security, but by honouring local knowledges and advancing an approach grounded in learning, attentiveness, and relational engagement. Instead of treating food scarcity as a problem to be solved in advance, we sought first to understand how local foodways operate and adapt in response to multiscalar pressures, including environmental change, urbanization, extractivism, and political-economic conditions determined by global market competition and war in the Middle East.

Moving away from externally imposed solutions and universal models, this book aims to make visible stories emerging from the pluriverse (Kothari et al., 2019): a multiplicity of worlds, practices, and ways of knowing that already exist and that challenge the hegemony of capitalist and colonial modernity (Quijano, 2007). In the current crisis of capitalism, reconnecting with practices that persist within capitalist ruins (Tsing, 2015) becomes a political necessity, as these practices cultivate forms of solidarity through which transformative social, ecological, and economic change can be collectively enacted (Ramcilovic-Suominen, 2025). The volume therefore engages with postcapitalist community economies (Gibson-Graham, 2006), practices of conviviality (Illich, 1973), and feminist scholarship on care as a gendered and political practice, enacted also through meal cultures (Harcourt, 2025).

Moreover, the book is indebted to Radical Food Geography, particularly as articulated by Hammelman et al. (2020). In line with their framework, the contributions engage with three interconnected elements of a radical food geographies praxis: critical theoretical engagement with power and structures of oppression, often drawing on local and non-white feminist scholarship; analysis grounded in specific geographic and ecological contexts; and research-action that involves collaboration with civil society beyond the immediate scope of the research project.

Importantly, these frameworks do not constitute abstract theories to be applied from above. Instead, they guide to engaging with very concrete practices of food production, harvesting, preservation, and meal sharing. These practices are not merely about nutrient intake or economic survival; they are relational spaces where sociality, affection, cultural transmission, and learning are continuously enacted. Shared meals, collective harvesting, and everyday food work become sites of informal pedagogy, where knowledge circulates through doing, observing, and caring-with others, human and nonhuman alike.

Engaging with community foodways, the book shows how immersive and participatory methodologies such as storytelling, making food together, visual practices, and community mapping, render food systems visible not merely as functional responses to material need, but as cultural ecologies of conviviality, care, and resilience. Often forged in tension with the homogenizing logics of industrial agriculture and globalized food regimes, which generate spatial pressure, environmental degradation, and political uncertainty, these ecologies persist through everyday practices of cooperation, improvisation, and resistance, enabling communities to creatively negotiate access to resources and sustain forms of collective life in contexts marked by profound inequality.

—

This is a small book. It does not claim to offer universal solutions or scalable models. Rather, it proposes a way of looking at the world grounded in close, relational research. The authors write from their experiences developed alongside communities. The knowledge presented here emerges from shared practices, long-term relationships, and mutual reflection. In this sense, the book can be read as a learning tool for students, scholars and practitioners interested in relational methodologies oriented towards postdevelopment futures.

While the second chapter offers explicit methodological reflections on relational, feminist, and postcolonial research practices, the chapters that follow present the empirical research conducted in specific urban and rural communities in Jordan and Morocco. The final chapter focuses on a crucial yet often overlooked phase of the research process: the reporting back of research findings to the communities involved. This took place during a moment of collective gathering, a local fish festival, where research results were shared and discussed in a communal setting.

In terms of empirical focus, while much of the existing literature privileges rural agricultural systems centred on subsistence and small-scale farming, this volume deliberately turns to less studied foodways, particularly artisanal fishing and urban food production. The case studies include Sidi Bounouar, a small village on the southern Atlantic coast of Morocco, where fishing is practiced by hand from rocky shores or with simple rods, as well as the coastal area of Tangier, where artisanal fishing has become increasingly marginalized amid urban waterfront transformations and inaccessibility of the coast. The book also engages with urban food practices in the rapidly expanding city of Amman, where the sprawl over formerly agricultural land has prompted urban groups to reclaim and cultivate a few remaining green spaces collectively. In parallel, residents have also formed initiatives aimed at enacting postcapitalist social relations through pedagogical engagements with food and food preservation practices, as alternatives to large-scale food distribution systems tied to capital-intensive economies and the political economy of war.

The research unfolded under critical political conditions. Fieldwork took place during an anomalous period marked by intense conflict in the Middle East. One field site initially planned in Lebanon became inaccessible, and even in areas without direct conflict, the pervasive atmosphere of war, together with our own presence as researchers affiliated with a European institution, shaped everyday life, relationships, and research practices. The authors' actions and engagements were inevitably shaped by this broader political environment, reinforcing the need for situated, responsive, and ethically attentive research. The absence of community members as co-authors reflects the nature of this writing as a side product of broader relational processes that materialize in diverse forms of action and collaboration extending beyond the production of academic texts.

Taken together, the chapters in this book invite readers to rethink food systems not as technical problems to be solved, but as relational worlds to be cared for. They suggest that attending to the everyday practices through which people grow, harvest, preserve, and share food can open up political imaginaries grounded in care, justice, and more liveable multispecies futures.

Acknowledgements This book emerged from the *Post-development Geographies of Local Food Systems: Community-based Networks Addressing Food Insecurity* project, whose funding was essential to make this research possible. Without this support, the work presented here could not have taken place. Together with the other authors of the chapters that make up this volume, I wish to express my deep gratitude to all those who participated in the project and with whom we built relationships of collaboration, dialogue, and mutual learning. Special thanks go to Chiara Rabbiosi, Ruth John, and Angela Kronenburg García, who worked on the parallel strand focusing on Maasai food communities in Tanzania. With them, we engaged in extensive discussions on the readings that inspired this work and on fieldwork methodologies, in an ongoing and generative exchange. We are particularly grateful to Chiara for her comments on single chapters' drafts. We would also like to thank Yafa el Masri and Mariasole Pepa, who initiated the idea for this research project and later pursued different paths. We are also grateful to all those who shared meals, time, and knowledge with us in Morocco—in particular the Amoud Association and the Sahel Aglou Women's Cooperative for Seafood Products in Sidi Bounouar, as well as ThinkTanger, the Donkey Museum, and fishers from *Bab el Bahr* in Tangier—and in Jordan, especially organizers and participants in Taghmees, Al Barakeh Wheat and Farfaheeneh in Amman. Further thanks go to Egidio Dansero and the Food Studies Lab for their insights on urban food networks; Elisa Bignante for discussions on the use of visual methods; Alessio Surian for the space offered to discuss group dynamics and research methodologies, and Massimiliano Tabusi, Gino Barsella, and Wendy Harcourt for the discussions held at the Italian Geographical Society Seminar on community-based foodways. Wendy Harcourt, in particular, has been a constant source of inspiration on feminist approaches. Many thanks to her for engaging with us on our project, and for providing thoughtful feedback since its earliest stages. We are also grateful to Will Lafleur for extensive conversations on sensuous research methods and for carefully reading and commenting on this volume. Finally, we would like to thank all those who shared their ideas and reflections during various workshops and conference sessions, particularly those sponsored by the Royal Geographical Society's Food Working Group and by Italian geographical associations.

References

Gibson-Graham, J. K. (2006). *A Postcapitalist politics*. University of Minnesota Press.

Hammelman, C., Reynolds, K., & Levkoe, C. Z. (2020). Toward a radical food geography praxis: Integrating theory, action, and geographic analysis in pursuit of more equitable and sustainable food systems. *Human Geography, 13*(3), 211–227.

Harcourt, W. (2025). *Conundrums of care: Feminist entanglements in critical development studies*. Bloomsbury.

Illich, I. (1973). *Tools for conviviality*. Harper & Row.

Kothari, A., Salleh, A., Escobar, A., Demaria, F., & Acosta, A. (2019). *Pluriverse: A post-development dictionary*. Tulika Books.

Mitchell, W. J. T. (2002). *Landscape and power*. University of Chicago Press.
Patel, R., & Moore, J. W. (2017). *A history of the world in seven cheap things: A guide to capitalism, nature, and the future of the planet*. University of California Press.
Quijano, A. (2007). Coloniality and modernity/rationality. *Cultural Studies, 21*(2–3), 168–178.
Ramcilovic-Suominen, S. (Ed.). (2025). *Socioecological transformations: Linking ideas with structures, personal with collective change*. Routledge.
Robbins, P. (2012). *Political ecology: A critical introduction* (2nd ed). John Wiley & Sons, Ltd.
Tsing, A. L. (2015). *The mushroom at the end of the world: On the possibility of life in capitalist ruins*. Princeton University Press.

CHAPTER 2

Relational Approaches to Food Research: Questions, Processes, Methods

Paola Minoia, Tamara Taher, Ginevra Montefusco, and Beatrice Ferlaino

Abstract As food studies increasingly embrace values of pluriversality, justice, and transformation, a gap remains in reflecting on the tensions that develop between neoliberal academic modes and times of knowledge production and the necessary attention to ethical integrity, epistemic diversity, and relational approaches. This chapter seeks to address that gap by navigating this contradiction and presenting a methodological

P. Minoia (✉) · T. Taher · B. Ferlaino
Department of Cultures, Politics and Society, University of Turin, Turin, Italy
e-mail: paola.minoia@unito.it

T. Taher
e-mail: tamara.taher@unito.it

B. Ferlaino
e-mail: beatrice.ferlaino@unito.it

G. Montefusco
Department of Historical and Geographical Sciences and the Ancient World (DiSSGeA), University of Padua, Padua, Italy
e-mail: ginevra.montefusco@phd.unipd.it

P. Minoia (ed.), *Foodways As Sites of Knowledge and Transformation*, Emplacing Food, https://doi.org/10.1007/978-3-032-28126-5_2

framework for studying food as situated, relational, spatial, and more-than-human practices.

We consider food, here, not only as a research topic but also as a methodological and epistemological lens for understanding relationships among humans, more-than-humans, and environments. After offering some key concepts and definitions in food studies, the chapter presents a set of embodied and relational research practices with the aim of exploring and outlining the epistemic opportunities inherent in them. Food is ultimately addressed in the chapter as a methodological entry point for analysing socioecological transformations, forms of social resistance and agency, experiences of being-in-common, practices of meaning-making and ways of re-imagining and reclaiming space and place. Finally, the chapter reflects on responsibility and accountability in relational research, and on how to make research processes and outcomes meaningful for and respectful of the priorities, times, and necessities of participants in them.

Keywords Relational food research · Situated reflexivity · Decolonial and feminist methodologies · More-than-human practices · Socioecological transformations

1 Food and Relationality, While Navigating Neoliberal Academia

What could be more visceral, embodied, and participatory than research grounded in food and conducted in close proximity to producing communities? Engaging with food across its entire cycle, from agro-silvo-pastoral and aquatic production and harvesting, to processing, distribution, consumption, and the practices through which food is cared for, honoured, and imbued with cultural, sometimes sacral value, situates research within material, affective, and relational worlds (Wilson, 2008). In this sense, food is not merely an object of inquiry but a methodological and epistemic medium through which knowledge is co-produced.

Participating in food-related practices entails entering spaces of convivial commoning, proximity, and co-presence (Harcourt, 2025). It allows research to move beyond extractive observation and beyond the final moment of meals consumption, towards an understanding of the social, ecological, and political challenges that have transformed production systems, territorial relations, and forms of subsistence. Through shared labour, everyday practices, meals, and rituals, research becomes an embodied process of learning-with and situated engagement. Approaching food through relational and grounded approaches foregrounds embodied and sensory knowledges, responsibility, ethics, and more-than-human entanglements (LaFleur, 2023). Honouring food through collective preparation, ritualised practices, or ethical attention to its trajectories, can be understood as an epistemic and political act: a way of recognising food as a nexus of land, labour, memory, cosmology, and power, and of resisting the commodification of life that characterise dominant food systems.

However, this type of research is hardly appreciated by funding agencies and academic research environments, which tend instead to privilege projects with objectives that are directly functional to increasing food production, optimising supply chains, enhancing consumption, or delivering scalable and measurable "solutions" to food-related problems. Research that unfolds through long-term engagement, relational accountability, and non-instrumental participation often appears inefficient, insufficiently innovative, or lacking in immediate impact according to dominant evaluative criteria. As a consequence, our work is often conducted within the very same neoliberal research environment it seeks to critically address, in an academic system marked by underfunded public research, competitive funding, short-term grants, productivist publication regimes, and audit-driven metrics of performance and impact. Research is increasingly embedded within international cooperation frameworks and policy-oriented programmes that prioritise scalability, impact, and economic growth. This context actively shapes the legitimacy of research questions and scientific methods, and the forms of knowledge produced and circulated. Therefore, even within critical orientations in our research, we are requested to produce project-based, rapidly circulating, and reportable knowledge aligned with imperatives of efficiency, and academic productivity. This raises a pressing question: how can we navigate this system without reproducing its epistemic, political, and material violence?

Food-based, community-engaged research offers a partial response to this dilemma. By insisting on relational, slow, and situated practices of knowing-with, it challenges extractivist epistemologies and neoliberal temporalities. Food functions as a vector of social relationality among humans and between human and non-human beings, and as a multiscalar lens that grounds research in embodied, situated experiences while connecting them to broader analytical scales. By weaving together economic, political, and symbolic dimensions, food renders social injustices visible through questions of access to resources and illuminates historical transformations and changing relationships with ecosystems (Hammelman et al., 2024). Taken together, these characteristics make food a valuable methodological entry point for remaining attentive to the key questions shaping research in contemporary academia.

A first set of issues concerns knowledge production itself. What kinds of knowledge are produced when research is organised around predefined objectives, timelines, and deliverables, often set before meaningful engagement with the field? Who has access to spaces of knowledge production? Without the co-production of knowledge that challenges globally defined research agendas and takes seriously situated, bottom-up experiences, the outcome is an ignorant form of knowledge: one that remains detached from lived realities, reproduces epistemic hierarchies, and fails to grasp the complexity of the interrelated social and ecological worlds it claims to represent. And crucially, whom is it produced for, and whom are we accountable to? Project-based research frequently aligns more closely with funders' expectations of delivering reports, indicators, and policy-oriented outputs, than with the needs and priorities of the populations involved. Outputs are often produced in English, circulated through academic journals, consultancy, or institutional reports, and closely tied to the operationalisation of "development" initiatives, risking opacity and limited relevance for local interlocutors. These dynamics raise critical questions that must be addressed when conducting research with communities, with whom it is instead fundamental to co-define goals that are meaningful, context-specific, and responsive to shared priorities.

Conducting research on food also requires a thorough awareness of the multiscalar relations that shape food production and access. Such awareness is crucial for navigating the field responsibly and for avoiding the exposure of local collaborators to political, social, or material risks, especially in contexts marked by conflict, surveillance, militarisation, environmental crises, and shifting regimes of control.

Such contexts of control often implicate research already in its preparatory phases, requiring engagement with authorities and the navigation of bureaucratic constraints, such as research permits or institutional authorisations. These interactions are not merely technical or administrative procedures but shape access to the field, the conditions under which the research is conducted, and the relationships established with those involved. This awareness may require adapting research questions, modifying or suspending fieldwork, or withholding materials to avoid exposing participants or researchers to risk. Although rarely visible in final outputs, these methodological and ethical decisions constitute a central dimension of a responsible research practice. Food can form part of strategies for accessing complex fields by enabling political dimensions to be approached in concrete and embodied ways, grounded in the situated perspectives that research participants identify as meaningful, rather than through abstract or predefined analytical frames. In this way, food-based research opens space for engaging with power through everyday practices and relations, offering an alternative to externally imposed readings while maintaining a degree of protection for both participants and researchers.

Conducting responsible, relational, and grounded research in a neoliberal academy requires continuous reflexivity on research agendas, methods, and practices in order to protect interlocutors, align projects with their priorities, and collectively define what must remain unsaid (in line with Glissant's (1990) notion of the 'right to opacity'). These negotiations are relational and shaped by the researcher's positionality. Because decolonial and feminist geography often focuses on marginalised groups or invisibilised spaces, knowledges, and practices, it demands heightened ethical attention and sustained reflexivity on how power operates within research relationships. A key risk lies in fixing communities into static identities, erasing internal diversity and obscuring differences related to gender, generation, class, or political positioning. Attending food access and exclusion, and how these dynamics are understood and reproduced, enables power to be analysed as relational, dynamic, and continuously reconfigured in everyday life.

Finally, a grounded decolonial and feminist research calls for an openness to pluriverses (Kothari et al., 2019) and to existing forms of kinship among humans and between human and non-human beings (Haraway, 2016). Food offers a privileged entry point to these relational entanglements, revealing interdependencies that may emerge when relations of trust are established with research participants. At the same time, it is required to recognise when such relations do not take shape—whether due to limited time or because the researcher's presence is neither necessary nor welcome. In many contexts, this reserve is rooted in historical experiences of colonial, patriarchal, or other forms of structural violence, and the difficulty of building 'familiar' relations must be acknowledged without essentialisation or romanticisation. Even under these conditions, food may offer alternative ways of creating contact, enabling the negotiation of proximity: depending on the context, it can provide access to political dimensions without requiring personal intimacy, or allow engagement at an individual scale without directly addressing broader political dynamics that may be uncomfortable for interlocutors. Decisions about with whom to work and which issues to address must therefore emerge through open negotiation, starting from what is already present and collectively shaping the form, distance, and dynamics of the research process.

In this chapter several concrete examples will be developed to explore how to navigate neoliberal academia while pursuing responsible, relational, and grounded research. This requires a critical and reflexive engagement with the contradictions that inhabit the research process, recognising methodology not as a neutral set of tools or techniques, but as a political, ethical, and relational space.

2 The Question of Time: Temporalities and Turbulences in Food Research

Researching food necessarily entails constantly relating to time. Whether we look at food chains, food systems on multiple scales, daily or seasonal practices, individual or collective meaning-making, food is experienced, made, shared, produced, consumed, wasted, and so on and so forth, in time. Capitalist production compresses and accelerates the temporalities and economies of food; grassroots and collective practices (such as festivals, celebrations, holidays, and community gatherings) expand the times of food sharing (Wilson, 2008); religious and spiritual practices and festivities slow down the relationship with food through fasting and food re-distribution (Hellman, 2008). The question of time is intertwined with the matter of food, and so with the practice of researching this specific subject matter. Even when analysing past processes and events, research is always happening and taking place in the present. It is therefore informed by the conditions of the latter, and carried out within a specific historical moment, territorial context, and social, political, and economic configurations. In the case of researching food, it is deeply intertwined with and affected by the processes informing it globally, regionally, and locally, and therefore cannot unfold without constantly interacting with them. This becomes particularly true and evident in moments of crisis, when movement and mobility cannot be taken for granted, and the connection between time and space is made most evident. Crises often reveal the relational and relative nature of the constructs of time and research, as well as of how food is made, in an entangled and globalised world.

In this sense, the times of the food we research and their plural rhythms, flows, lengths, and frames, are a context of research as much as place, space and territory are. Several orders of time need to be held into consideration when researching food: the times of academic research and higher education, and the times of the food processes researched. The times of knowing in academic research itself are multiple (Smith, 2015): there is the time designed and imagined in the research project, the time informed by the work conditions of an increasingly neoliberal, fast-running, academia (highly influenced by funding mechanisms and reporting necessities) and higher education (Felt, 2016). On the other hand, there are the temporalities of the world the researcher interacts with: the time— or lack of it— of the people the researcher aims to engage with, the social and political times of different societies involved in the research process, the present historical time of global changes and regional crises, the contextual times of food production, consumption, sharing, etc. All such times run at different paces, affecting the research process and the researcher's ability to access and construct both theoretical analysis and fieldwork.

It is well established across disciplines (including the social and hard sciences) that research does not neatly follow the schedules and timelines of its initial design, nor does it appear as unitary and monolithic (Felt, 2024). Temporal turbulences, however, are not simply an inconvenient or unexpected element in processes of learning, and of building and producing knowledge. They are a constitutive part of the subject matter of research (McLeod, 2017) and can thus be approached as "conceptual and analytical tools" (Hohenthal & Minoia, 2022, 852), i.e. as disruptive events that reveal the non-linearity of the processes that the research analyses, as well as the non-orderly character of research itself.

This is relevant for research on food in times of global crises when inequalities in access to food, starvation and hunger are weaponised in conflicts (such as in the Gaza Strip, Sudan, Yemen, etc.), or can result from climate and ecological crises that result from global extractive capitalism, food production and resource management policies across decades (for example: in fires, floods, natural disasters, crop diseases, or even pandemics and epidemics). In a world that is becoming more politically and ecologically unstable (leading to a resurgence of discourses of closure, armed conflict, and racism), researching food can become a space of learning about all that affects it and is affected by it, on a multi-scale that goes from individual and local experiences to international processes. This ultimately entails an attention to the multiple temporalities of food production, consumption, movement, practices, and knowledges globally and locally, as well as an awareness of how all that influences the temporalities of research itself. Put more simply, we cannot pretend to study food outside of the world that produces and consumes it, or outside of our own bodies interacting with it daily. A relational approach to researching food is fundamental in this direction, as it takes account of the ever-dynamic nature of eco-cultural transformations revolving around and through food.

The Covid-19 global crisis in 2020, conflicts across the world (which keep on evolving in the present) and changing international politics have concrete effects on what doing geographic and social research about food looks like in our time. Concretely, adopting such a relational approach involves keeping into the analysis time as a dimension that is produced and constructed: it depends on the conditions of the researcher and her/his interlocutors. Each relationship is always contextual, each time is always specific to how the relationship evolves. Concentrated and short, or prolonged periods of field research function based on the subjects involved. Sometimes, occasions such as festivals, workshops, specific events, or activities offer a convergence of actors and processes that allows the observation of multiple dynamics and the dialogue with different figures. Other times, long and slow times of permanence within a research context give space to understanding the tensions and staying with the cracks produced by both structural processes and attempts of re-imagining the world. Each research temporality is tied to how the researcher moves along with the community and inside the context of research. What is relevant and important is how such time is valued and made meaning of inside the research, and how the researcher practices such time respectfully and by taking care of doing no harm to others and to the contexts she/he learns about. Finally, an attention to the relationship between multiple timelines and temporalities and research calls for re-thinking and re-imagining methodologies and access points, as well as the very definition of what context and field to carry out research in, meaningfully and ethically.

3 Key Concepts in Relational Food Research

Food is a complex subject that escapes disciplinary divides. Food studies have long intersected multiple domains, such as environment, space, gender, race, and health, as well as a wide range of approaches and disciplines, including anthropology, political economy, ecology, and development studies, among many others (Holloway et al., 2025; Kneafsey et al., 2021; Parker et al., 2019; Slocum & Saldanha, 2016). Food studies also embody transdisciplinarity, as they draw on concepts and methods that cross academic boundaries, bridging together fields such as the visual and performative arts alongside natural sciences, including agronomy, marine biology, and ethology. From a geographical perspective, the plurality of food offers a generative space for both theory and practice, particularly in research that explores the relationship between place, ecologies, bodies, and communities. At the same time, this complexity poses significant challenges, as researching food requires holding together the situated dimensions of culture, space, economy, politics, and materiality through which food is produced and circulated (Goodman, 2016).

To face this tension, some geographers have conceptualised food as *more-than-food* (*ibid.*), thereby questioning the traditional divide between food production and food consumption as main focuses of the field debate. A *more-than-food* perspective positions food in an in-between space, where what food is—and how it comes to matter—is shaped through embodied relations and ordinary practices. This debate has stimulated the development of new, transformative methodologies: according to Goodman, food must be "*more-than-followed*", drawing from the well-known *follow-the-thing* approach proposed by Ian Cook (2004). In other words, food should be researched across its chains of value and production, while adopting theoretical frameworks and methodological lenses that can attune with plural food worlds. Such an approach entails attending to food's materiality, spatiality, and temporality—and, in doing so, to ecological erosion, the transformation of territories, and the industrialisation of ecological rhythms. Alongside this, it calls for a reflexive and situated involvement with the people and communities who negotiate—and, at times, resist—these challenges, in both ordinary and creative ways.

Approaching this framework raises both ontological and epistemological questions: What is food, and through which channels do we come to know it? How is knowledge generated, developed, and shared through food, beyond its production and consumption?

Moragues-Faus and Marsden (2017), in their food research agenda grounded in political ecology, identify three focal points for further explorations of food-ecology-society entanglements. First, unravelling places as scales where socioecological relationships are materially and symbolically manifested. Second, tracing the trans-scalar nature of political processes that shape food systems and related injustices. And third, attuning food as a generative practice of meaning-making and social change.

This section offers a range of food-related concepts which can be mobilised in geographical food research in order to unpack these layers from a relational, situated, and embodied perspective.

3.1 Emplacement

Food and place are co-produced in material and symbolic ways. The notion of *baladi* foodways offers a useful lens through which to understand this interdependency. *Baladi* is a polysemantic term in Levantine Arabic that intertwines geographical meanings of village, land, and locality with the affective and political meaning of homeland. *Baladi* foodways are therefore emplaced, namely shaped by local, traditional, and artisanal practices that are adaptive to the specific ecologies where food is produced, as a result of centuries of socioecological mediation. Importantly, *baladi* does not refer only to physical location. It also embodies the quality of food that exceeds nutritional values, including cultural compatibility with the place and communities inhabiting it, while also functioning as a political claim to land and food sovereignty. At the same time, neither place nor *baladi* foodways are a-temporal. Like other local, indigenous, artisanal food practices in other cultural contexts, *baladi* foodways are forged by centuries of circulations and encounters across cultures, within uneven geometries of power (Nasser, 2024).

The emplaced nature of food thus extends beyond the sites of production and consumption to encompass the spatial, ecological, and symbolic dimensions that constitute place in a relational perspective (Garden et al., 2021). Place is shaped by everyday practices, trans-scalar circulations, ecological encounters, and narratives and imaginaries. Attending to food practices as sites where such narratives are entangled gives room to unrecognised, marginalised, and colonised voices and their epistemic power. Place, in sum, must be conceived as relational, with relationality foregrounding extended networks of both injustice and change.

3.2 *Trans-Scalarity and Mobility*

Food lacks the stillness typically attributed to material things. It traverses bodies, nourishes memory, and draws a sense of place, actively mediating relationships between ecologies and bodies. Food is inherently mobile and transformative: it changes forms as it is cultivated, processed, transported, cooked, and ingested. This essential mobility calls for a trans-scalar analytical gaze, capable of following food across multiple trajectories, material transformations, and social meanings.

What we call food is a complex matter emerging from interaction between organic and inorganic elements within specific ecologies—such as bacteria, soil, water, infrastructures, and chemicals. These interactions are regulated by production techniques, economic policies, and technologies, and give shape to the food product which enters local and global market circuits in the form of a commodity. More than being consumed, food is experienced in places and through relationships. It creates spaces of encounter and conflict, tracing paths through the guts, domestic spaces, city streets, and global political agendas.

In these movements—both physical mobilities and shifting imaginaries—scale becomes a crucial analytical concern. Ignoring locally embedded networks of production, exchange, and consumption, and advising their replacement with externally introduced practices reliant on imported inputs or products, contributes to the erosion of biological and cultural diversity and fosters relations of dependency from external and globalised markets that often translate into vulnerability and impoverishment (Patel & Moore, 2017). Thus, attending to scale allows us to trace food miles and supply chains, but also to reveal how ecological transformations, political mobilisations (such as boycotts), and economic structures shape the places and practices of food. At the same time, trans-scalar gazes foreground intimate relations at the scale of the body, home, and kitchens, which are key for understanding affective politics, practices of care, and everyday forms of resignification, resistance, and transformation. Trans-scalarity thus politicises intimate and domestic spaces, situating them within broader processes of power and change.

3.3 Socioecological and Spatial Change

Adopting a trans-scalar approach to food in place makes visible the extent to which movement, transformation, and interdependence are constitutive of food practices. From a relational perspective, social and ecological dynamics co-produce one another, participating in the transformation of foodways and their spaces.

On the one hand, this perspective draws attention to how social dynamics emerge in response to ecological and spatial transformations impacting food. Communities engaged in farming or fishing, for example, constantly face climate crises, land or coastal privatisation, and resource extraction in their territories. These processes, which reshape both the ecologies and spaces where food is produced and consumed, often generate adaptive, collaborative, and resilient strategies, as well as political struggles and collective mobilisation to sustain food practices, as exemplified by movements claiming food sovereignty or agroecology. On the other hand, a socioecological lens also attends to how specific ecologies actively participate in shaping the material and symbolic dimensions of food practices. However, the spatial dimension of this co-production is often overlooked, especially in urban context. Coastal communities offer a clear example of how food practices are shaped by ecological and spatial relations with the sea, through temporalities, rituals, and techniques that adapt to marine rhythms and spaces. Yet, urban coasts are increasingly exposed to real estate speculation and privatisation, not only eroding marine ecologies but also neglecting access to key spaces for artisanal fishing. While the impact of neoliberal urban planning on urban natures erosion (both land and oceans) is often acknowledged, less relevance is given to the spatial accessibility of sensitive spaces for fishing, foraging, and gardening, which are crucial for sustaining urban food practices and the socioecological relations embodied in these places. Food practices, thus, can be understood as sites where socioecological relations are stabilised, contested, or reworked over time and space.

Socioecological and spatial changes, thus, are central to food practices and raise critical questions about who controls resources, who bears the costs of environmental change, and whose food systems are rendered viable or marginal. Political ecology offers a critical framework for analysing food systems by situating food production, access, and consumption within relations of power, inequality, and contested resource governance (Holt-Giménez & Shattuck, 2011). From this perspective, focusing on communities of practice entails following the political, ecological, and economic processes through which foodways are governed, resisted, and reconfigured in everyday life. A key dimension of these transformations concerns how communities respond to these shifts through practices of resilience, resistance, and reinvention.

3.4 Community

The concept of community has been subject to substantial critique within post-development and feminist studies. As Morton (2000) observes, the *C-word* has often functioned as a camouflage behind which governments withdraw from economic responsibility, while promoting moralistic conservative values that intensify social control. Post-capitalist and feminist scholarship, in this direction, calls for liberating "community" from its "traditional" framing as sameness, shared identity, or harmonious cohesion. Instead, community is rethought in terms of interdependence and difference, grounded in the social character of economic life (Gibson-Graham, 2006). This shift moves the analytical focus from fixed notions of culture and identity towards practices of being-in-common, becoming, and commoning.

In *Pluriverse: A Post-development dictionary* (Kothari et al., 2019), community is described as a term that "pushes to the foreground the profound interdependence humans have with each other and with non-human others" (p. 129). This definition provides a relational and socioecological lens for approaching food communities as assemblages that include non-humans not merely as resources to be consumed, but as active co-producers of knowledge and sociality (Tsing, 2014). From a post-development perspective, food communities can be approached along two complementary directions. On the one hand, more-than-human communities may embody an alternative model to dominant and extractive approaches in food production, starting from situated knowledges and adaptive practices. On the other hand, this literature cautions against how development intervention—once a community is defined as a homogeneous stakeholder—reshapes local contexts, influences values and priorities, and enhances a romantic and coherent representation aligned with funding dynamics while fostering internal conflict. Feminist approaches bring further complexity for a critical reading of community by warning against the mobilisation of Western-centric gender categories in the understanding of food practices and roles. At the same time, this scholarship foregrounds care, difference, and everyday practices as central to the political and economic life of communities.

The concept of "conviviality" offers, in this sense, an interesting and stimulating perspective on how to think of community, raising crucial questions about dominant paradigms of development and progress. In *Tools for Conviviality* (1973) Ivan Illich defines it as the opposite of industrial productivity, describing it as "autonomous and creative intercourse among persons, and the intercourse of persons with their environment", grounded in individual freedom realised through interdependence. From relational and post-development perspectives, conviviality enables a re-thinking of foodways beyond extractive and growth-oriented models of food security. It foregrounds practices of mutuality, care, and shared responsibility between humans and their ecologies, while resisting the reduction of social relations to instrumental or market logics and while acknowledging conflict, plurality, and asymmetries of power as constitutive of social life.

4 Methods

How can we practice food research in meaningful, honest, and respectful ways? Approaching research through food brings attention to everyday practices and material engagements, offering an entry point for participatory and creative methodologies. Food-based research invites methods that move beyond interviews and textual analysis to include visual approaches, collaborative mapping, and shared practices of production, processing, cooking, eating, and care. These practices enable forms of co-presence and learning-with that are grounded in sensory experience, embodied knowledge, and relational accountability. Food enables the co-construction of research methods that emerge through doing, rather than being designed in advance, challenging extractivist and purely discursive approaches to knowledge production. It also opens up accessible and situated ways of sharing research that extend beyond academic texts, such as collective meals, visual artefacts, and storytelling. The following section introduces some of these methodological approaches, not as an exhaustive toolkit, but as a situated selection grounded in our research practice and expertise. While food-based research opens up a wide range of creative forms of expression, many of these require artistic or technical skills, as well as broader collaborations, that extend beyond our own competencies. Therefore, we focus here on methods that are accessible, adaptable, and ethically grounded, and that can be mobilised across diverse research contexts.

4.1 Multisensory Spatial Observation

Spatial observation is a first entry point for understanding how food practices are embedded in space. Observing food pathways (where food is produced, transformed, exchanged, cooked, and consumed) requires attention not only to visual signs, but to multisensory and embodied experiences (Pink, 2015; Sutton, 2010). While the visual gaze is often considered dominant in spatial analysis, other senses such as smell, sound, touch, and taste are equally important for acquiring knowledge. Food spaces, such as agricultural fields, urban markets, collective kitchens, streets, and waterways, are experienced through the whole body, including visceral or "gut" feelings that connect bodily sensations, memory, and cognition.

Orientation relies on multiple sensory systems and is shaped by social, political, and environmental contexts. Body position and movement are informed by sensory perception and elaborated through orientation, making orientation not only a cognitive mapping of space but also a lived and embodied experience. For example, moving through a market involves navigating smells, sounds, crowd density, and social norms; working in a field entails bodily attunement to soil, weather, and tools; participating in a collective kitchen requires negotiating shared rhythms, roles, and forms of care.

Observation, also while walking or participating in food-related activities, can focus on objects (infrastructures, tools, food items) as well as on phenomena and actions (exchanges, gestures, routines, conflicts). Conducting observations at different times of the day, week, or year reveals how food spaces are differently used and accessed by diverse human and non-human actors, including different generations, genders, ethnicities, abilities, and animals. For instance, remaining unsold food in a market may become accessible to some marginalised groups only at specific moments, such as during market closure, when informal or organised re-distribution takes place; similarly, a field or a kitchen may involve gendered or age-specific roles.

Observation may be serendipitous or guided by specific research questions; participant or non-participant; systematic or open-ended; conducted alone or together with local residents, producers, or cooks.

Mapping and counter-mapping can then be used to document these experiences, mark significant places and routes, and reflect on accessibility, organisation, history, and the sensory and affective dimensions of food spaces.

4.2 Walk Along

The walk-along method is a qualitative research approach that combines movement through space with interviews and other forms of spatial exploration, enabling the production of embodied and contextual knowledge of space. It allows researchers to engage with spaces through what research participants themselves consider important, valuable, and meaningful, fostering a situated and grounded perspective on the processes shaping the studied context.

Moving together with people living the studied context makes it possible not only to access how spaces are understood and narrated, but also to apprehend dimensions that would otherwise remain at a purely discursive level: the difficulty of traversing certain paths, their duration, the surrounding landscapes and its transformations, the meanings attributed to these changes, and the exploration of Indigenous ontologies through attention to everyday interconnections with the environment. These are just some of the elements that tend to escape purely verbal exchanges. Attending interconnections with the non-human world (Springgay & Truman, 2018), and sustaining counter-narratives over certain contexts (Powell, 2024), is a key dimension of this method. The walk-along approach enables the researcher's direct participation in the everyday life of the studied context, while also allowing participants to define—quite literally—where research goes. Following food pathways, exploring food spaces, and understanding where food is produced, transformed, consumed, and how it circulates allow not only a deeper understanding of food systems, but also illuminate meanings, practices of care, and convivial ways of being together that remain invisible without embodied and shared engagement (Morgan, 2023). At the same time, this approach makes visible the agrarian and aquatic landscapes through which food moves, as well as processes of environmental degradation and regeneration— including urban expansion, soil and forest erosion, and contamination, which are deeply entangled with everyday food practices.

To conduct a relational and grounded research, an important aspect of this method lies in its openness to contextual variation. Walking carries different meanings across contexts: it may constitute a privilege or a necessity, and walk-along practices must therefore attend to local specificities, embodied differences, and uneven conditions of mobility (Ernsten & Shepherd, 2024). Focusing on these differences in the study of food can illuminate inequalities and power dynamics that often remain inaccessible through interviews or methods that do not involve embodied participation in food-related practices.

4.3 Community Mapping and Counter-Mapping

Community mapping and counter-mapping are participatory approaches that produce bottom-up representations of space, foregrounding local knowledges, experiences, and priorities often marginalised in official cartographies and highlighting "the social meaning and historical political development of spatial attributes" (Hohenthal et al., 2017, p. 234). While they share an attention to situated perspectives, they differ in their political orientation and degree of explicit contestation. Community mapping commonly aims to collectively represent lived spaces and may seek visibility, dialogue, or planning without necessarily positioning itself in direct opposition to dominant narratives (Parker, 2006). Counter-mapping, by contrast, more consciously operates within logics of resistance and dissent, explicitly challenging hegemonic representations, exposing power relations, and supporting claims against dispossession, exclusion, or erasure (Halder & Collective Orangotango, 2019).

These distinctions are contingent rather than fixed. The actors involved, the political context, and the purposes for which maps are produced shape both the method and the scope of the mapping activity. The creation of these alternative representations often works through counter-narratives that may be openly dissident, subtly oppositional, or silenced and ignored; in some contexts, the political value of mapping lies precisely in making visible what is framed as 'empty' or insignificant. Importantly, dissent can take multiple forms: in certain settings, simply proposing an alternative narrative already constitutes a political act. Community mapping initiatives may also become counter-mapping practices through their use, for instance when grassroots maps are mobilised to legitimise bottom-up claims in engagements with institutions. The methods through which such narratives are constructed are diverse, focused on participation, and often combined, ranging from focus groups, walk-along methods, and mental or sensory maps to participatory GIS and other cartographic tools, allowing grounded mapping to remain flexible, situated, and responsive to context.

Food can play a crucial role within these practices (Wight, 2019). As a vector connecting multiple scales, food enables mappings to link embodied experiences to broader environmental, political, and economic dynamics. Food-based mapping can function as a tool for understanding, monitoring, and contesting spatial transformations from below, particularly in relation to environmental change and questions of justice tied to access to land, freshwater, marine environments, and other resources. By acknowledging the spatial dimensions of environmental degradation, resource extraction, and uneven exposure to risk, such practices support situated forms of observation and control that are often absent from official representations. At the same time, political, symbolic, and collective dimensions make visible relations of access, exclusion, care, and dispossession, grounding counter-narratives in everyday life.

4.4 *Storytelling*

Storytelling is a qualitative research method that centres a practice that permeates human societies and daily lives across the world: telling tales (Rooney et al., 2016). Everyday individual and collective stories have not been deemed as methodologically sound and epistemologically relevant for a long time by academic disciplines. It was through critical approaches across anthropology (Falconi & Graber, 2019), sociology (Erasga, 2010), and human geography (Cameron, 2012) that stories and the practice of recounting them have been progressively acknowledged as spaces of knowing and being. While this recognition might seem new in some social sciences, storytelling has been crucial in other epistemologies and worldviews in indigenous societies (Iseke, 2013) and colonised contexts defending their worlds from material and epistemic destruction and injustice. This is relevant when reflecting on how to carry out storytelling methods in different contexts: it is important to understand how people the researcher interacts with have often already long traditions, wide and shared practices, and even significant theorisation about the epistemic and social value of telling stories.

In the context of the Arab region, for example, there has been significant reflection, not only in the academic sphere but also in cultural spaces outside of it, about the practice of *haki* ("talking") and how it represents a socially embedded way of learning, knowing, and transmitting knowledge and testimony (Hamzeh et al., 2020). Palestinian learning theorist Munir Fasheh widely reflects on *qusas* (stories) and *hakaya* (tales) as radical forms of learning. In his work across decades, he explores how telling *hakaya* and *qusas* constitutes an altogether different epistemology than modern theory-making (Sukarieh, 2019). In the past decades, people across the region have engaged in reclaiming time and space for collective storytelling and traditional forms of listening to, performing, singing, dancing stories. Whether fantastic tales or lived personal experiences, the stories recounted in such spaces often open windows on the collective socio-cultural, political, and economic processes that have shaped their societies and communities across time. As food represents a socio-cultural space and not only a material practice of nourishment, stories are often shared around it and can thus represent significant ways of knowing and crossing through it. People's stories of food can narrate a lot about their collective social, cultural, economic, and political histories. The Gaza Strip's cuisine, for example, has historically been influenced by the people's access, or lack of it, to specific kinds of foods informed by occupation, siege, repeated military operations and aggressions across the years, and starvation in the past years. Recipes and stories of how food is made in such a context speak of wider and layered processes (El-Haddad & Schmitt, 2013). Similarly, migrants' stories and memories of food and of feelings of *ghurba* ('estrangement') in Turin's market at Porta Palazzo recount personal experiences of collective mobility, borders, transnational communities, social change, and economic relations (Proglio, 2025).

4.5 *Cooking, Eating, Doing Together*

Logocentric methodologies are often insufficient to entirely represent the complexity of real-life interactions. Methodologies involving experiential moments, sensitive exploration practices and embodied methods have developed more recently to enrich both research methodologies and epistemologies of food (Wilbur & Gibbs, 2018), especially in connection to the growth of post-human approaches in the social sciences and food studies (Elton, 2019). Acknowledging the pedagogic and epistemic nature of the body and of doing things with food (such as producing, conserving, eating, cooking, fasting, offering and sharing, etc.) has been increasingly received in the field of studies as it allows to acknowledge what goes unnoticed in solely narrative approaches describing relationships with and of food. Abbots argues, indeed, that eating is an epistemically significant act, in which "the affects and effects of food on the body are fully realized" (Abbots, 2015, 14). She underscores how through eating "the body engages with both the materiality and the meaning of food", underlining how "'glossing over' eating risks overlooking such moments of engagement and the micro-practices of contestation and acceptance inherent within them" (*Ibid.*, 14). Like eating, cooking also involves senses that are usually overshadowed in approaches that centre intellectualisation as a mode of knowing and seeing as a way of learning. Smell, touch, sound, and taste can collect information differently, including in implicit ways that speak of complex and layered collective and personal processes that take place through and around food (MacDougall, 2021).

Embodied, sensitive, and experiential methods allow relational research on food as they entail a full involvement of the researcher and not simply the role of a "participant observer". Making food together, and/or any shared food practice, with others can also fill in the gaps between different spoken languages, and diverse personal or collective histories, often offering space to share unexpected aspects of one's memories, thoughts, and feelings. In this sense, it can become a precious space of learning and relating to one another, where the researcher does not only direct their own gaze upon others but is more fully present with them and gazed upon by them as well.

4.6 Photography: Documenting, Participating, Sharing

Within food studies, photography is widely adopted to investigate the processual and phenomenological dimensions of making and sharing food. It offers a "way in" to everyday food practices—such as cooking, farming, and foraging—by enabling researchers to look "along with" rather than "at" those engaged in them (Wills et al., 2016). This method has been widely adopted in studies focused on the domestic sphere, allowing for in-depth observation of the ordinary and familiar food routines. Photographic methods are frequently combined with mobile research practices, such as go-alongs and walking interviews, supporting the documentation of what participants share about food practices while capturing atmospheres and sensory dimensions of significant places.

At the same time, photography is not a neutral recording of reality, but a situated and negotiated practice, historically engaged with colonial and extractivist modes of knowledge production (Hall, 2009). The power dynamics embedded in photography have generated ethical debates concerning both production and analysis of visual data, calling for sustained reflexivity around positionality in the photographic process (Rose, 2022). This growing awareness can be traced in two directions. The first concerns more-than-representational approaches which interpret images beyond the subject depicted or the visual output per se. This scholarship emphasises the situated and affective nature of photography considering the phases of negotiation, sharing and circulation of images as a key part of the knowledge production process, deeply embedded in the gaze of both the photographer and the viewer. Within food research, photography can thus be understood as a situated practice which does not simply represent food and food-related activities but actively generates processes of meaningful sharing, especially in research contexts where linguistic barriers need to be navigated creatively. In these cases, sharing pictures as a final outcome of research in dedicated restitution events– such as photographic exhibitions within community gatherings– provides a powerful way of returning the research process to participants and broader communities, making it more horizontal and accessible. The meaning of pictures, in this key, emerges not just from the represented subject as seen by the researcher, but also from the impression and reaction of people while engaging with them.

Participative approaches, on the other hand, focus on the need of sharing the epistemological power of photography with research participants. Within this framework, photo-elicitation and photo-voice represent two key participatory methods. Photo-elicitation—first introduced by Collier in 1957—uses photographs as prompts during interviews or focus groups to stimulate discussions about memories and the emotional dimensions evoked by the picture, provided both by the researchers and the participants. Photo-elicitation has been widely adopted to stimulate discussion about food choices, routines, and perceptions, especially with young people and women (Vassenden & Jonvik, 2022); with a focus on ecological relations and natural resources (Bignante, 2010; Johnson, 2018); as well as a learning tool for geography classrooms (Kurtz & Wood, 2014).

Photo-voice, informed by Freirian critical pedagogy, extends this approach by placing the camera in participants' hands, proposing photography as a tool for collective reflection and action-oriented research. Photo-voice activities are usually conducted with a group of participants which share a concern on a specific topic - for instance, inaccessibility of local agricultural products in a certain neighbourhood. Through three main phases, namely training, individual image production, and collective discussion of visual materials, photo-voice allows diverse perspectives to be articulated and negotiated collectively, fostering shared narratives and dialogue oriented towards concrete pathways for change (Wang & Burris, 1997). Photo-voice has been adopted in food research for participative planning and action-oriented research, especially with marginalised groups or vulnerable contexts (Diez et al., 2017).

4.7 Video-Making: Building Multisensorial and Shared Narratives

Over the last fifteen years, attention to film has intensified within geographical scholarship, in parallel with the growing centrality of qualitative research and embodied forms of inquiry grounded in practice, experience, and performative creativity. Film adds layers of complexity to still images: sound and movement make video extremely immersive in conveying atmospheres and sense of place. Imagine how the representation of a street market radically changes when voices of people selling and bargaining can be heard, or how the coastline gets livelier when the movement of waves, seagulls, and boats passing by can be seen. Beyond their uses as sources, documentation, or supporting materials within the research process, films have their own language, capable of producing situated and meaningful knowledge of spatial relations and processes (Bignante & Memoli, 2025).

While implementing video-making as a method, all the phases of filming, editing, and discussing footage need to be accounted as crucial epistemological moments. In other words, the way in which we frame a subject, we record soundscapes, and we imagine the storyboard before editing the different sequences together, shapes the multisensorial narrative we want to convey. This process can be developed both collaboratively on the field or by the researcher alone. Collaborative filmmaking has been widely mobilised within critical and creative geographies as a tool for building shared narratives around social issues, including gender-based violence, ecological struggles, and questions of justice and care (Jacobs & Salimbeni, 2024).

Within food geographies, video-making is particularly valuable for its capacity to engage with temporality, movement, and sound, namely key dimensions through which foodways are experienced. Thanks to the communicative power of moving images, participatory video-making is often a space for radical counter-narrative and encounter between food communities and researchers, supporting their struggles throughout the production and dissemination phases (Levkoe et al., 2023). Videos can also be proposed to participants as field diaries as a strategy to give back a daily narrative of food practices and to archive the many overlooked activities surrounding the preparation of a meal. Given the sensitive nature of video-making, encompassing issues such as the visibility of people and potentially contested places, it must be reflexive of contextual dynamics. In practice, filming and shooting can be perceived as dangerous in both private and public spaces. To adopt these methods ethically, careful attention must be paid to potential harm to participants and discomfort, beyond just obtaining informed consent. Sometimes, the most ethical option is to leave certain subjects, places, and voices opaque. Alternative editing strategies, such as focusing on soundscapes, layering pictures with drawings, or creating collages, can be used as techniques to creatively share stories and represent contested matters while maintaining an ethical approach.

4.8 Events and Fairs as Spaces for Collaborative Research

Research work, especially when conducted by researchers who live outside the communities being studied, often interacts with local communities in ways that can be intrusive. Entering people's spaces and rhythms of everyday life may require residents to accommodate external needs out of hospitality, including requests to participate in interviews, focus groups, or meetings. This can place uneven burdens on participants and reproduce extractive dynamics, even when research intentions are critical or collaborative (Smith, 1999).

In some contexts, however, it may be easier and more appropriate to identify shared spaces of encounter. Local events, fairs, and festivities, particularly those celebrating local food and products, can offer such opportunities. Ethnographic research on festivals and fairs has shown how these temporary gatherings can function as meaningful field sites, where food communities take shape through active social relations, cultural practices, and forms of expertise that become visible through concentrated interactions and performances (Amo, 2023). Food fairs often bring together a dense constellation of actors: producers of regional and traditional foods, local institutions and agencies supporting production, associations, and service providers. These spaces are characterised by a general openness to exchange, explanation, and dialogue, as exhibitors are encouraged to present their activities, expertise, and needs.

Compared to markets, fairs can be especially informative, as participants are motivated to explain how products are made, where they are processed, how medicinal plants or spices are used, and how local dishes are prepared. Public discussions, roundtables, and informal conversations frequently take place alongside exhibitions, allowing researchers to listen, observe, and engage without imposing structured research settings. Walking through stands, initiating conversations, or offering contributions and services can become part of a reciprocal research practice.

Local events can also provide opportunities to share research findings together with local experts and practitioners, and to receive meaningful feedback from an interested audience. Creating such spaces requires preparation, collaboration with event organisers, and a willingness to contribute beyond research needs, working alongside local associations and actors. In this way, events become sites for relational, situated, and collaborative ethnographic research.

5 Conclusions

The intent of this chapter has been to offer key concepts and reflections on fieldwork methodologies for researching food practices within feminist, decolonial, relational and grounded research. Across the sections, food has been approached as both an object of study and a methodological and relational medium through which research can be grounded in everyday practices, embodied experiences, and socioecological entanglements. The chapter argues that food (in its several phases of production, processing, circulation, preparation, sharing, and consumption) can be a fertile entry point to create relationships and engage simultaneously with intimate spaces and with broader political-economic and environmental dynamics. This trans-scalar capacity of food allows for an approach that roots the analysis of structural issues in the everyday dimension of people participating in the research, enabling attention to access, exclusion, care, and dispossession without abstracting these dynamics from lived experience.

A central contribution of the chapter lies in its discussion of relational tools and approaches to research. While relationality is often invoked as an ethical ideal, this chapter has argued for a political understanding of relationality that resists romanticisation. Relationships are shaped by asymmetries of power, uneven burdens of participation, and historical experiences of colonialism, development, and extraction. Conducting research responsibly and ethically, therefore requires reflexivity on positionality, on how relationships are formed and maintained over time, and on who carries the labour of hospitality, translation, and care. Relational methods are not inherently emancipatory; their ethical and political significance depends on how they are practiced and negotiated.

The methodological practices discussed in the chapter are presented as situated and adaptable approaches. Their common strength lies in their capacity to move beyond extractive research practices, enabling co-presence, learning-with, and the co-construction of knowledge through doing. These methods draw attention to sensory, embodied, and affective dimensions of food practices, making visible knowledges, forms of care, and agencies inscribed in everyday practices that can remain overlooked or misunderstood if considered from a non-situated perspective.

Importantly, these tools also offer different degrees of proximity and exposure. The chapter has highlighted how methodological choices (where to work, with whom, how visibly, and through which practices) are inseparable from questions of care towards participants and accountability of the researcher. Food-based methods such as those presented here can allow analyses that are grounded in everyday practices, and that are able to explore the complex and nuanced ways through which political, economic, and socio-cultural processes unfold, with an attention not to exploit and expose participants. At the same time, ethically-grounded approaches insist on recognising when proximity is neither possible nor appropriate, and on respecting refusal, silence, and opacity as meaningful positions rather than as methodological failures.

Acknowledging the importance of respecting our interlocutor's desires, this chapter has delved into the complexity of pursuing responsible relational research in a neoliberal academy which does not always recognise the importance of placing boundaries and respecting case-specific times—which does not always align with the expectations of neoliberal productivism and project-oriented rationality. Radical approaches—such as the decolonial, feminist, post-development, post-capitalism, and political ecology—have provided an important analytical grounding for navigating these tensions. By situating food practices within contested resource governance and relations of power, they enable research to move beyond technical accounts of food systems, towards an analysis of socioecological change, inequality, and struggle. At the same time, feminist and more-than-human approaches complicate simplified narratives of community resilience by emphasising the plurality of actors and strategies, internal conflict, and differentiation, as well as care practices and interdependence. In this sense, this chapter mainly proposes to overcome fixed or romantic representations of the people we work with, often framed as 'the local community', while considering them instead as non-homogeneous and dynamic, always responding to environmental and economic processes in diverse context-specific forms (resistance, adaptation, reinvention) shaped by gender, generation, class, and political positioning.

Finally, the chapter has reflected on responsibility and the question of restitution. Giving back cannot be standardised or reduced to predefined outputs. It must be understood as a situated and negotiated practice that takes different forms depending on context, needs, and relationships. It may involve sharing findings in accessible ways or supporting existing initiatives. In some contexts, responsibility may instead require limiting circulation and visibility to protect those involved. What matters is not the specific form that giving back takes, but the commitment to accountability, reciprocity, and care through which contributions are negotiated and enacted.

In conclusion, the chapter has argued that methodology is not a neutral set of tools, but a political, ethical, and relational space. The food-based methods discussed offer concrete possibilities for conducting feminist and decolonial geographic research that remains grounded in everyday life while engaging critically with power, inequality, and socioecological transformation. Navigating neoliberal academia from this perspective does not mean finding a position outside its constraints, but rather continuously negotiating its contradictions: resisting extractive logics, valuing time and relationships, and insisting that research remains accountable to the communities and ecologies through which it becomes possible.

References

Abbots, E. (2015). *The Agency of Eating. Mediation, Food and the Body*. Bloomsbury Academic.

Amo, E. (2023). The Slow Food Movement and the Terra-Madre project: Food Sovereignty And Translocal Assemblages. *Globalizations, 20*(4), 644–660. https://doi.org/10.1080/14747731.2022.2149158

Bignante, E. (2010). The Use of Photo-Elicitation in Field Research. Exploring Maasai Representations and Use of Natural Resources. *EchoGéo* (11) https://doi.org/10.4000/echogeo.11622

Bignante, E., & Memoli, M. (2025). *Geografia Filmica: Il Film Geografico e il Video Partecipativo*. Mimesis.

Cameron, E. (2012). New Geographies of Story and Storytelling. *Progress in Human Geography, 36*(5), 573–592. https://doi.org/10.1177/0309132511435000 (Original work published 2012).

Collier, J. (1957). Photography in Anthropology: A Report on Two Experiments. *American Anthropologist, 59*(5), 843–859.

Cook, I. (2004). Follow the Thing: Papaya. *Antipode, 36*(4), 642–664.

Díez, J., Conde, P., Sandin, M., Urtasun, M., López, R., Carrero, J. L., Franco, M., et al. (2017). Understanding the Local Food Environment: A Participatory Photovoice Project in a Low-Income Area in Madrid, Spain. *Health & Place, 43*, 95–103.

El-Haddad, L., & Schmitt, M. (2013). *The Gaza Kitchen: A Palestinian Culinary Journey*. Just World Books.

Elton, S. (2019). Posthumanism Invited to Dinner: Exploring the Potential of a More-Than-Human Perspective in Food Studies. *Gastronomica, 19*(2), 6–15. https://www.jstor.org/stable/26854597

Erasga, D. S. (2010). When Story Becomes Theory: Storytelling as Sociological Theorizing. *Asia-Pacific Social Science Review*.

Ernsten, C., & Shepherd, N. (2024) Walking as Embodied Research: Drift, Pause, Indirection. Routledge.

Falconi, E. A., & Graber, K. E. (Eds.). (2019). *Storytelling as Narrative Practice. Ethnographic Approaches to the Tales We Tell*. Brill.

Felt, U. (2016). The Temporal Choreographies of Participation: Thinking Innovation and Society from a Time-Sensitive Perspective. In J. Chilvers & M. Kearnes (Eds.), *Remaking Participation: Science, Environment and Emergent Publics* (pp. 178–198). Routledge.

Felt, U. (2024). Time and Temporalities in Academic Research. In Felt, U. & Irwin, A. (Eds), *Sociology, Social Policy and Education* (pp. 318–327). Elgar Encyclopedia of Science and Technology Studies. https://doi.org/10.4337/9781800377998.ch33

Garden, E., Walters, V. M., & Chamberlain, K. (2021). Reasserting Food in Place: The Case of Kai Whau. *Social & Cultural Geography, 22*(7), 936–955. https://doi.org/10.1080/14649365.2019.1672776

Gibson-Graham, J. K. (2006). *A Postcapitalist Politics*. University of Minnesota Press.

Glissant, É. (1990). *Poétique de la relation*. Poétique II. Gallimard.

Goodman, M. K. (2016). Food Geographies I: Relational Foodscapes and the Busy-Ness of Being more-than-Food. *Progress in Human Geography, 40*(2), 257–266.

Halder, S., & Orangotango, K. (Eds.). (2019). *This is Not an Atlas: A Global Collection of Counter-Cartographies*. Unchanged second edition. Transcript (Social and cultural geography, Vol. 26).

Hall, T. (2009). The Camera Never Lies? Photographic Research Methods in Human Geography. *Journal of Geography in Higher Education, 33*(3), 453–462.

Hammelman, C., Levkoe, C. Z., & Reynolds, K. (2024). *Radical food geographies: power, knowledge and resistance*. Bristol University Press (Food and society).

Hamzeh, M., Carmona, J. F., Sánchez, Ma. E. H., Bernal, D. D., & Bejarano, C. (2020). Haki/pláticas ~ testimonios/shahadat: Arabyya feminista decolonial praxis. *Journal of Curriculum and Pedagogy, 17*(3), 250–265. https://doi.org/10.1080/15505170.2020.1786750

Haraway, D. (2016). Staying with the Trouble: Making Kin in the Chthulucene. Duke University Press (Experimental futures. Technological lives, scientific arts, anthropological voices).

Harcourt, W. (2025). *Conundrums of Care: Feminist Entanglements in Critical Development Studies*. Bloomsbury.

Hellman, J. (2008). The Significance of Eating During Ramadan: Consumption and Exchange of Food in a Village in West Java. *Food and Foodways, 16*(3), 201–226. https://doi.org/10.1080/07409710802304176

Hohenthal, J., & Minoia, P. (2022). Territorial and Mobility Justice for Indigenous Youth: Accessing Education in Ecuadorian Amazonia. *Mobilities, 17*(6), 850–866. https://doi.org/10.1080/17450101.2021.1987154

Hohenthal, J., Minoia, P., & Pellikka, P. (2017). Mapping Meaning: Critical Cartographies for Participatory Water Management in Taita Hills Kenya. *The Professional Geographer, 69*(3), 383–395. https://doi.org/10.1080/00330124.2016.1237294

Holloway, L., Goodman, M. K., Maye, D., Kneafsey, M., Sexton, A. E., & Moragues-Faus, A. (Eds.). (2025). *Elgar Encyclopedia of Food and Society*. Edward Elgar Publishing.

Holt-Giménez, E., & Shattuck, A. (2011). Food Crises, Food Regimes and Food Movements: Rumblings of Reform or Tides of Transformation? *The Journal of Peasant Studies, 38*(1), 109–144.

Illich, I. (1973). *Tools for Conviviality* (1st ed.). Harper & Row.

Iseke, J. (2013). Indigenous Storytelling as Research. *International Review of Qualitative Research, 6* (4, Winter), 559–577.

Johnson, L. B. (2018). Becoming 'Enchanted' in Agro-Food Spaces: Engaging Relational Frameworks and Photo Elicitation with Farm Tour Experiences. *Gender, Place & Culture, 25*(11), 1646–1671. https://doi.org/10.1080/0966369X.2018.1541867

Kneafsey, M., Maye, D., Holloway, L., & Goodman, M. K. (2021). *Geographies of Food: An Introduction*. Bloomsbury Publishing.

Kothari, A., Salleh, A., Escobar, A., Demaria, F., & Acosta, A. (Eds.). (2019). *Pluriverse: A Post-Development Dictionary*. Tulika Books.

Kurtz, H. E., & Wood, J. (2014). Stone Soup: Photo-Elicitation as a Learning Tool in the Food Geography Classroom. *Journal of Geography in Higher Education, 38*(4), 546–556. https://doi.org/10.1080/03098265.2014.958657

LaFleur, W. (2023). Storying Sensuous Atmospheres of Peaches and Wheat: Towards New Horizons of Imagination and Possibility for Sustainable Food Systems. *Ethnologia Fennica, 49*(2), 66–100. Article 3. https://doi.org/10.23991/ef.v49i2.112979

Levkoe, C., Lowitt, K., Furlotte, S., & Sayers, D. (2023). Advancing Radical Food Geographies Praxis through Participatory Film: Reflections from an Indigenous-Settler Food Sovereignty Collaboration. *ACME, 22*(1), 762–790. https://doi.org/10.7202/1098035ar

MacDougall, S. (2021). Embodying Levantine Cooking in East Amman, Jordan. In: Gaul, G. A., Auman Pitts, G. & Valosik, V. (Eds.), *Making Levantine Cuisine. Modern Foodways of the Eastern Mediterranean* (pp. pp. 153–169). University of Texas Press.

McLeod, J. (2017). Marking Time, Making Methods: Temporality and Untimely Dilemmas in the Sociology of Youth and Educational Change. *British Journal of Sociology of Education, 38*(1), 13–25. https://doi.org/10.1080/01425692.2016.1254541

Moragues-Faus, A., & Marsden, T. (2017). The Political Ecology of Food: Carving 'Spaces of Possibility' in a New Research Agenda. *Journal of Rural Studies, 55*, 275–288.

Morgan, C. B. (2023). Studying Food Systems as Embedded, Sensory Phenomena. *Frontiers in Sustainable Food Systems, 7*, 1040965. https://doi.org/10.3389/fsufs.2023.1040965

Morton, T. (2000, November 4). Beware the C-word. *Spectrum, Sydney Morning Herald*, 1, 10–11

Nasser, C. D. (2024). Discussing Food and Foodways: No Better Time Than the Present. *Jerusalem Quarterly, 98*, 7.

O'Connell, R. (2013). The Use of Visual Methods with Children in a Mixed Methods Study of Family Food Practices. *International Journal of Social Research Methodology, 16*(1), 31–46. https://doi.org/10.1080/13645579.2011.647517

Parker, B. (2006). Constructing Community Through Maps? Power and Praxis in Community Mapping. *The Professional Geographer, 58*(4), 470–484. https://doi.org/10.1111/j.1467-9272.2006.00583.x

Patel, R., & Moore, J. (2017). *A History of the World in Seven Cheap Things: A Guide to Capitalism, Nature, and the Future of the Planet*. University of California Press.

Parker, B., Brady, J., Power, E., & Belyea, S. (Eds.). (2019). *Feminist Food Studies.* Canadian Scholars' Press.

Pink, S. (2012). *Situating Everyday Life: Practices and Places*. Sage.

Pink, S. (2015). *Doing Sensory Ethnography* (2nd ed.). Sage.

Powell, K. (2024). Walking the Map. *International Journal of Multicultural Education, 26*(1), 8–30.

Proglio, G. (2025). *Porta Palazzo. Storia orale e sensoriale del mercato più grande d'Europa*. Donzelli Editore.

Rooney, T., Lawlor, K., & Rohan, E. (2016). Telling Tales: Storytelling as a Methodological Approach in Research. *The Electronic Journal of Business Research Methods (EJBRM), 14*(2), 147–156, www.ejbrm.com

Rose, G. (2022). *Visual Methodologies: An Introduction to Researching with Visual Materials* (5th ed.). Sage.

Slocum, R., & Saldanha, A. (Eds.). (2016). *Geographies of Race and Food: Fields, Bodies, Markets*. Routledge.

Smith, S. (2015). Multiple Temporalities of Knowing in Academic Research. *Social Science Information, 54*(2), 149–176. https://doi.org/10.1177/0539018414566421 (Original work published 2015).

Smith, L. T. (1999). *Decolonizing Methodologies: Research and Indigenous Peoples*. University of Otago Press.

Springgay, S., & Truman, S. E. (2018). *Walking Methodologies in a More-than-Human World: WalkingLab*. Routledge (Routledge advances in research methods, 24).

Sukarieh, M. (2019). Decolonizing Education, a View from Palestine: An Interview with Munir Fasheh. *International Studies in Sociology of Education, 28*(2), 186–199. https://doi.org/10.1080/09620214.2019.1601584

Sutton, D. E. (2010). Food and the Senses. *Annual Review Anthropology, 39*, 209–223. https://doi.org/10.1146/annurev.anthro.012809.104957

Tsing, A. (2014). More-than-Human Sociality. A Call for Critical Description. In *Anthropology and Nature* (pp. 38–54). Routledge.

Vassenden, A., & Jonvik, M. (2022). Photo Elicitation and the Sociology of Taste: A Review of the Field, Empirical Illustrations, Arguments for A "Return to Photography." *The Sociological Quarterly, 63*(1), 154–174. https://doi.org/10.1080/00380253.2020.1834464

Wang, C., & Burris, M. A. (1997). Photovoice: Concept, Methodology, and Use for Participatory Needs Assessment. *Health Education & Behavior, 24*, 369–387.

Wight, A. (2019). *Food Mapping as Community-Based Participatory Research*. In Proceedings of the 2019 AERA Annual Meeting. 2019 AERA Annual Meeting, AERA. https://doi.org/10.3102/1431779.

Wilbur, A., & Gibbs, L. (2018). 'Try It, It's Like Chocolate': Embodied Methods Reveal Food Politics. *Social & Cultural Geography, 21*(2), 265–284. https://doi.org/10.1080/14649365.2018.1489976

Wills, W. J., Dickinson, A. M., Meah, A., & Short, F. (2016). Reflections on the Use of Visual Methods in a Qualitative Study of Domestic Kitchen Practices. *Sociology, 50*(3), 470–485.

Wilson, S. (2008). *Research Is Ceremony: Indigenous Research Methods*. Fernwood Publishing.

CHAPTER 3

Making Food and Meaning Together: Spaces and Temporalities of Being-in-Common in West Amman

Tamara Taher

Abstract Amman has witnessed rapid urbanization in the past decades. Urban rhythms and spaces, and processes of food globalization, have increasingly detached the city dwellers from knowledges and practices in relationship with the land and seasonal food processes. At the same time, in a present where starvation is used as a genocidal settler colonial weapon in the Gaza Strip, food has become central in boycott practices by Jordanian citizens. Questions of food sovereignty, security and economic dependency have been increasingly articulated, and food has often become a space linking solidarity with Palestine to questions of environmental and ecological crises in Jordan. Within such context, this research interacts with three spaces in West Amman reclaiming local food practices, as well as land-based epistemologies and pedagogies, and asks: how has the neoliberal urban development of the city informed the food practices and knowledges of West Amman's citizens in time? And what do these three spaces practice, in the here and now of their city's

T. Taher (✉)
Department of Cultures, Politics and Society, University of Turin, Turin, Italy
e-mail: tamara.taher@unito.it

P. Minoia (ed.), *Foodways As Sites of Knowledge and Transformation*, Emplacing Food, https://doi.org/10.1007/978-3-032-28126-5_3

contradictions, in terms of their relationship with food, land and community? The chapter traces links between the pedagogic efforts carried out by these subjects and post-capitalist challenges of detaching from, questioning, or performing disruptions of "development"-centred transactions and worldviews.

Keywords Post-development · Temporality · Food · Land · Conviviality · Amman

1 Food in Context, Entangled in Neoliberal Urbanism, Aid Dependency and Regional Crisis

"There is nothing else to do here but go to restaurants!", a frustrated friend visiting Amman said to me one day as we sat in a café in Jabal Al-Weibdeh, the increasingly gentrified historical neighbourhood hosting Amman's cultural and artistic scene, as well as many foreign students and "expats" working in international organizations (Albaik, 2023). In conversations with friends and MENA studies colleagues who have visited the city in the past decade, Amman is often perceived and portrayed as offering less cultural life and intellectual spaces than cities like Beirut, Damascus, Cairo and Jerusalem.

These accounts inadvertently reproduce a long-standing perception of Amman's "inherent lack of urbanity" among inhabitants and visitors alike (Shami, 1996, 2007). Shami problematizes the orientalist assumptions underlying gazes that seek the exoticism of the "Islamic city", a model based on European and Western descriptions of specific contexts (Fez, Aleppo, Damascus) and generalized on all Arab cities. At the same time, she has asked: "what is it that Amman lacks in order to qualify, in the eyes of its inhabitants, as a city?" (Shami, 2007). This question has been articulated in different ways also by Daher (2008) and Bustani (2024), reflecting how Ammanis perceive their city as struggling with affirming its specific cultural identity. Indeed, its population has developed through multiple waves of migration in the past decades. Yet, it is this plurality that defines, today, a modern Ammani identity reclaimed more actively (Daher, 2008), calling on inhabitants to listen, feel and observe the city and their presence in it more attentively (Bustani, 2024).

Since the early-2000s Amman has undergone a process of urban development that has attempted to enhance its role on multiple levels. The city has seen an increase in private investments, both foreign and local, and in global capital flow aimed at developing it through urban regeneration (Hourani, 2014) and tourism (Doan, 2006). The latter's growth, contributing 14% to Jordanian GDP, spurred investment in Jordan's infrastructure (Yousef, 2024), producing an expansion in services, markets, hotels and restaurants in the city in the past three decades (Abu Aqiqah & Al-rfou', 2010). Notable examples of neoliberal urban regeneration include the Abdali Project and Jordan Gate (Bustani, 2023a; Daher, 2013), both situated in West Amman (the wealthier side of the city), as well as the Greater Amman Municipality's regeneration project of Rainbow Street (Daher, 2011; Page, 2013).

These changes have contributed in different measures to shaping Amman's present-day foodscapes and foodways. International fast-food and local restaurants, supermarkets and food services have occupied an increasingly central role in the city's life amid the development of neoliberal spaces of consumption (shopping malls) and dynamics of urban sprawl, environmental change—particularly water scarcity and agricultural land depletion (Al-Johari & Sharaf, 2024)—and socio-cultural transformations (Schwedler, 2010).

Since the early-2000s, several national strategies have envisioned Amman's growth, from the *Amman 2025* Master plan (Beauregard & Marpillero-Colomina, 2010) to the National Green Growth Plan (Ministry of Environment, 2017) to the 2025 Planning and Zoning Regulations (Jordan News Agency, 2025). International and regional funding has significantly influenced these processes and broader Jordanian society. Aid contributes to a big extent to the economy, and affects the work of neoliberal civil society organizations receiving funding from international organizations and donor governments. Within such conditions, food security and agriculture receive 13.2% of foreign aid (Al-Assaf, 2025).

The relationship between aid, a neoliberally designed civil society and food production and consumption has come to the fore of collective attention with the Zionist aggression on the Gaza Strip since October 2023, and the ongoing genocide of the Palestinian people (Human Rights Council, 2024), which has spurred worldwide protests, including in Amman and other Jordanian cities. Throughout the time in which I was carrying out research, famine had been unfolding in Gaza for months.

The UN's World Food Programme declared its food stocks in the Strip depleted (World Food Programme, 2025) and the Gaza Humanitarian Foundation acted through security contractors in collaboration with the occupation's military as another death trap disguised in humanitarianism (Al-Jazeera, 2025a) before its operations were suspended in November 2025 (Al-Jazeera, 2025b). Warnings of mass starvation have been issued repeatedly since October 2023: from the Gaza Soup Kitchen and local farming initiatives to UNRWA, urgent calls to open crossings and allow food trucks have gone unheeded amid continued imperialist intervention in the region and military and economic support to the violence on Gaza.

The weaponization of hunger in Gaza has drawn regional attention to long-standing questions of food security and sovereignty, that had already become relevant during the Covid-19 pandemic and on other occasions, such as in the Jordanian popular campaign to end the Jordan-Israel gas deal. Since the signing of a letter of intent in 2014, and throughout implementation in the following years, the "*Ghaz al-ʿaduw ihtilal*" ("The enemy's gas is occupation") campaign has contested how the deal affects Jordanian sovereignty by impoverishing the country, depleting its natural resources and obstructing investment into renewable energy projects (Bustani, 2025). Since October 2023, the acceleration of colonial-imperialist violence in Palestine and the region has generated popular initiatives supporting Palestinian *sumud* ("steadfastness"), such as the "Revive Gaza's Farmland" by the Arab Group for the Protection of Nature, the "*Al-Hamla al-sha'biyya li-da'm sumud Ghazza*" (the "Popular Campaign to Support the *Sumud* of Gaza") and other Boycott, Divestment and Sanctions (BDS) campaigns. Boycotts of international fast-food chains and supermarkets, like Carrefour, cited for supporting the Zionist military, gained traction. Consequently, according to the Jordan BDS movement, the French supermarket officially closed its branches in Jordan in February 2025 (BNC, 2025).

All such efforts, accompanied by calls not to sign the energy-for-water deal between Jordan and Israel (Al-Jazeera, 2023), have generated growing discussion around solidarity and support to the Palestinian people, and on Jordanian food security and sovereignty (Tammey For Youth, 2024). Local products and restaurants replacing international options, as well as lists and apps of products-to-boycott, have proliferated as people considered their daily food choices as acts of solidarity and a way of rejecting economic complicity. At the same time, a growing desire for emancipation from international aid dependency has developed on several levels, from local organizations to cultural spaces to food production. Yet, the lack of meaningful local and regional alternatives to international development funding and foreign aid triggers frustration and debate on how to sustain boycott practices with longer-term strategies. The Trump administration's disruption of USAID (which flooded Jordan's economy for decades) in 2025 has inevitably generated crisis in a highly donor-dependent civil society (Mbah et al., 2025).

In this context, looking at spaces in the city rejecting international funding and expressing grounded and locally owned discourses and practices of material and epistemic re-appropriation can be meaningful. Their experiences, while constrained, articulate questions of detachment from and disruption of the development discourse (Escobar, 1995). Navigating the entire space of activities, restaurants and groups working on food in the city is wider than the scope of this chapter and requires a critical outlook on food narratives and practices by connecting them to economic and material relations and structures. Here, the analysis is directed at exploring three specific experiences in West Amman, brought together in a constellation of the present: *Taghmees*, *Al-Barakeh Wheat* and *Farfaheeneh*.

Each relates to food as a space of multiple entanglements—produced, consumed, shared, made meaning of and known within material and symbolic relations. Since 2012, Taghmees defines itself as a "social kitchen" and learning space, where cooking and eating together are practiced as embodied epistemological and methodological means of learning and making meaning. Founded in 2020, Al Barakeh Wheat works on producing *baladi* wheat and re-introducing local flours and grains in the Jordanian market. By doing so, it also emphasizes the collective reappropriation of knowledge about food and land. Finally, Farfaheeneh, a mobile social space, invites seasonal gatherings to re-learn food storing and production practices that are part of collective food memories across Bilad al-Sham. These experiences actively speak of reclaiming food, relationships with the land, local epistemologies and pedagogies from within the space of the neoliberal city, therefore countering, in different ways, development-based economic and epistemic worldviews, and attempting to imagine and practice liberated words and food.

2 Conviviality Vs. Productivity: Moving toward Epistemic and Material Emancipation in Urban Food-Land Relations

Food studies have explored the multiple lenses through which food can be approached and understood. Economic, social, political, ecological and cultural forces shape not only how food is produced, moves and reaches—or fails to reach—consumers in a globalized capitalist world; but also how people interact with food, make meaning of it and experience it in social and environmental relationships (Counihan et al., 2019) and in their bodies (Abbots, 2015). Social lives, collective stories and changing food practices— cooking, eating, selling, producing, boycotting, learning and so on—all participate in defining food as both a relational space and an agent. Abbots (2015) addresses, for example, the political dynamics and power relations of eating, and asks: "Can eating contest [...] political relations and produce new knowledge?' 'Can agency be located in the matter of food itself?' 'Is agency to be found in the interactions between eaters, food and knowledge-(makers)?' [...] 'In what ways can eating be an expression of agency?'" (Abbots, 2015, p. 1).

These questions are relevant as I reflect on how I usually attempt to respond to the frustrated visitor I mentioned above. How does food, in its urban hegemonic and subaltern configurations, discourses and practices (in the relationship between its socio-cultural meanings and material production and consumption), speak of the social fabric of this city? How does it embody the stances and relationships of its inhabitants with the city, the country, the region and the world? And, in the experiences of the people I have engaged with, how does reclaiming material and epistemic food practices speak to the present moment and to the historical processes leading to it?

Concepts such as "foodways" and "foodscapes" have articulated a layered gaze on food that "considers the social contexts that make food and make food meaningful—spanning from fields to markets to kitchens, factory spaces and restaurants» (Gaul & Auman Pitts, 2021, p. 2). Most recently, the Radical Food Geographies approach has emphasized the relation between power, knowledge and food, presenting itself as a "world-making project" committed to exposing the interrelated nature of hunger, geography and systemic injustices (Hammelman et al., 2024), and offering critical tools for understanding the structural roots of power relations shaping food systems in a present built on the expansion of neoliberal capitalism and deeply embedded in colonial and racial logics.

In a modern, rapidly growing and densely populated city like Amman, the food system must be understood in relationship to the territorial and environmental changes, as well as the economic transformations and policies adopted, across the past decades. Political economy and ecology perspectives are essential to understanding the production of Amman's urban space (Lefebvre, 2004) in modern Jordanian history, and how its transformations have shaped the inhabitants' relationship with land, resources and food. The city's environmental and territorial changes in less than a century have been characterized by a swift growth of the built environment at the expense of agricultural land and natural resources, including the coverage of the water stream that once flowed between Ras al-Ein ("head of the spring") and Ein Ghazal (Gharaibeh et al., 2019, p. 5). In contrast to the present of the city, the depiction of Amman in the Umayyad period offered by Gharaibeh et al. reflects a radically different landscape:

> A collection of natural springs from Ras Al-Ein (formerly known as Wadi Abdoun) and the rainwater runoff from the western mountains maintained the stream flow. The fresh waters supported agricultural lands and provided the basic everyday needs of the settlers. Amman was famous in the Umayyad era (636–750 AD) for the cultivation of grain and animal husbandry [9–11]. [...] The importance of Amman as a trading centre and an exchange market has been further enhanced by the presence of several flourmills (run mills), turned by the flowing stream waters. (Gharaibeh et al., 2019, p. 2)

This account foregrounds the ecological and economic centrality of water and agriculture in Amman's past, in stark contrast with its present condition as a water-scarce, densely urbanized territory. The stream was progressively covered by the municipality through development plans aimed at managing the 1960s and 1970s population growth, driven by regional crisis following the Palestinian "*Naksa*" (the 1967 Six Day War). The expansion of the built environment over agricultural lands westward continued in the following decades (Abu al Haija & Potter, 2013, pp. 2–3) as urban sprawl unfolded in an unregulated manner, frequently linked to further regional crises, such as the Iraq war in 2003 and the Syrian conflict (post-2011), spurring additional waves of migration. Amman's urban population grew from 1.7 to 4 million between 2004 and 2020 (Abdeljawad & Nagy, 2021, p. 3004). Al-Johari and Sharaf estimate that if trends persist, agricultural land in the city will be depleted by 2053 (Al-Johari & Sharaf, 2024, p. 211).

These material and demographic transformations intersect with the broader decline of agriculture in Jordan, a process shaped by liberalization policies over the past decades (Al Akayleh, 2013). In this regard, Hopma has explored through a Gramscian lens, centred on the analysis of the dialectical relationship between socio-cultural, political and material dimensions, how trade-based food provisioning impacts Jordan's agriculture and food system, and how farmers and their representatives in the Ministry of Agriculture have contested the open-market paradigm produced by globalization (Hopma, 2017, pp. 137–147). On the other hand, Bustani points to how liberalization policies developed in connection to the International Monetary Fund's loan conditionalities led to popular protest in the late 1980s, most notably in "*ahdath al-khubz*" (the "bread riots") in 1996, when wheat subsidies were lifted, leading to a sharp increase in the price of bread—a decision that was partly reversed in response to popular unrest (Bustani, 2023b, p. 37). On another level, a growing literature on energy geopolitics in the region (Schuetze & Hussein, 2023) also critically assesses the effects of regional tensions on Jordan's access to water, gas and renewable energy, and their multilayered impact on environmental change, migrant and local labour, food sovereignty and energy policies.

Alongside these transformations, the early-2000s marked a pivotal shift in Amman's urban trajectory. The city has undergone neoliberal urban restructuring (Daher, 2013), with emphasis on projects aimed at transforming its image, especially the western side. While Daher takes a more critical stance on urban neoliberal policies, Schwedler examines how neoliberal cosmopolitanism influences Amman's inhabitants' aspirations and consumption practices. Both approach urban neoliberalism not as a single monolithic framework of theoretical assumptions and policy models, but in its specific, contextual manifestations—shaped by the individuals and social groups involved in processes of urban change and transformation. Daher calls for a complex analysis reconciling neo-Marxist understandings of hegemony and poststructuralist ideas of discourse and governmentality to investigate urban neoliberalization (Daher, 2013, pp. 101–102).

This chapter argues that combining the here-outlined perspectives can support researching the city's food system. Specific planning and economic policies have not only produced neoliberal urban spaces (as in the proliferation of restaurants, fast-food outlets, supermarkets and shopping malls) but also significantly influenced the food practices of Ammanis across spatial, economic and social divides. Yet city dwellers are not merely passive recipients of neoliberal policies; rather, they actively engage with these dynamics, often aspiring to participate in the city's emerging "cosmopolitan" spaces (Schwedler, 2010).

Building on critical understandings of urban space, some scholars and practitioners propose alternative planning paradigms, such as self-planned communities (Kably, 2015), that centre the relationship between planning processes, community needs and environmental sustainability. On the other hand, literature exploring the relationship between time and space—for example through "rhythms" (Lefebvre, 2004), melodies and territories (Brighenti & Kärrholm, 2018)—focuses on the coexistence of multiple temporalities, both hegemonic and subaltern, inhabiting the city. Lefebvre elaborates "rhythmanalysis" by looking at layers and transformative relations between times: from regulated time, governed by rational laws (numerical, quantitative and qualitative), to embodied, lived, carnal rhythms in the human being (respiration, the heart, hunger and thirst, etc.) (Lefebvre, 2004, p. 9).

Analysing the diverse rhythms of food practices and knowledges enables research to look at both food and the city through the lens of the "pluriverse" (Escobar, 2018), a concept popularized in post-development and decolonial theories that does not simply imply multiplicity but gestures toward diverse ways of life, knowledge systems and ontologies, specifically those marginalized by colonial-capitalist structures. By envisioning a "world where many worlds can exist", the "pluriverse" reopens the horizon that capitalism's worldview closes, disavowing any alternatives to itself (Fisher, 2009).

Ivan Illich's critique of capitalist productivity informing modern societies, and his concept of "conviviality" (Illich, 1973), offer here stimulating lenses to rethink the relationship between people and their environments in material, ontological, temporal and epistemic terms. Far from representations of conviviality as mere inclusion in sharing food, Illich's theorization articulates an alternative to the productivity paradigm governing all aspects of modern capitalist societies, including the urban space. A "convivial society" counters the ideal of "progress" at the heart of development discourse, designating "the opposite of industrial productivity". It is defined as the "autonomous and creative intercourse among persons, and the intercourse of persons with their environment [...] in contrast with the conditioned response of persons to the demands made upon them by others, and by a man-made environment" (Illich, 1973). Illich places conviviality within interdependence, as both a material and ethical order countering the imperatives of capitalist productivity, not negating the value of technology in bettering collective and individual lives but stressing an approach that conceives of tools "to work with" rather than "work for" the human.

Illich's thought has inspired post-development theory, and, in the context of Jordan and Palestine, it has been approached by Munir Fasheh, a Palestinian learning thinker who has developed his *ta'allum* (learning) praxis in radical closeness to personal and collective stories and experiences. *Ahali* ("kin" being its closest English equivalent) and *hikayat* ("oral stories") are central in Fasheh's epistemology. His perspective is a theory of detachment from modernity's "plastic language" and of rooting back meanings and learning in community (*mujaawarah*, an Arabic word for "neighbouring") and in the relationship with material and civilizational soils (*turba*) (Fasheh, 2021). Taghmees, a group in Amman that works on reclaiming knowledges rooted in their local social fabric and environmental context, has interacted with Fasheh for many years through the founders' understanding of the practice at the heart of their name, emphasizing the word's meaning ("dipping bread") and its verb's root (pointing to "immersion"). Knowledge and meaning, here, are closely intertwined with the community and food practices that the group has developed across time, in connection to their critical questions emerging from the need to redefine their understanding of their environments and ecologies. While not articulating theory in academic spaces, Taghmees has long dialogued with indigenous, decolonial and "global South" contexts on many occasions. I bring their experience and voice into theory analysis to practice, in the writing of this chapter, a relational and grounded approach to ideas.

Finally, the courage and willingness to imagine and practice worlds not aligned with modern imperatives of "progress" and reclaiming other understandings of "abundance"/"scarcity", calls for approaches that attempt to stay in the contradictions of the capitalist hegemonic order. Post-capitalist theory (Gibson-Graham, 2006) addresses the question of liberating economic transactions from capitalist hegemony and practicing other ways of being-in-common while not escaping the materiality of a world tangled in capitalist structures. By centring "being-in-common" instead of "community", Gibson-Graham focuses on the ever-contextual practice of coming together to think and practice alternative relations and worlds based on interdependence and complexity instead of sameness, linearity, positivity, fullness, commonness.

In the context of Amman, characterized by aid dependency and development conditionalities, neoliberal privatization and space production— and within global crises of weaponized hunger—what does it mean to hope for, and attempt to practice, a different world in the here and now?

3 The Researcher within the Research: Building Knowledge from Presence at the Thresholds

Conducting research in one of my own worlds—in a city where I live today, where family stories stretch back decades—is profoundly different than doing research elsewhere. I carry personal memories and inherited narratives that shape how I see and move through this place. This familiarity carries the joy of rediscovering what I thought I knew, but also responsibilities and difficulties. These challenges have been especially relevant in a present marked by the ongoing genocide in Gaza, and considering my research's critical lens on development, aid dependency and food practices and knowledges in this city.

Post-development critique has informed the very structure and interactions between me and the people I wanted to engage with. This close entanglement, and lack of separation, between research and life is both a theoretical and methodological question: it opens epistemic and embodied anxieties (Newton, 2022) that need to be taken seriously. In this, I have resonated with indigenous research theories underscoring the local researcher's dilemmas with her own community and the violence embedded in the colonial function of academic research (Tuhiwai Smith, 2012) in contexts subject to capitalist extractivism, neo-colonialism and imperialism. In particular, with the bleeding wound of the genocide in Gaza produced by Zionist settler colonialism—and the critique expressed by many in Jordanian society toward international donors seen as complicit—approaching fellow Ammanis as a researcher working in an Italian EU-funded research project wasn't comfortable both for my own self and for some of my interlocutors.

I did embark on this research project in a very sensitive time, few months after the beginning of the aggression on the Gaza Strip. In a context of regional upheaval, I was a "researcher" who wanted to study food practices and spaces resisting development logic and elaborating other ways of being together, but my project was very much part of the post-Covid-19 European development discourse and connected to an international actor widely criticized locally. This is where my standing at the threshold between societies, my multiple positions—as an Italian early career scholar attempting to critically navigate the neoliberal university; an Ammani citizen; and a Palestinian dedicating all her work to Palestine—first clashed and then spoke to one another. The unease is never resolved, but this multiplicity allowed me to conceptualize my work as a practice of my theoretical stance: I came to an understanding that, from a position of liminality, I was navigating a hegemonic-pedagogic relationship (Gramsci, 1971) to bring forward, imagine and articulate other possible worlds than the development-based and "progress"-oriented capitalist order.

Who I was mattered to my interlocutors. The groups I have interacted with refuse funding or collaboration with certain international actors, and, for some of them, with the university itself as an institution of modern education. However, in our conversations—sometimes uncomfortable—whenever I clearly shared how the research was funded, I was met with the recognition of who I was as a person and as a researcher/learner: a member of this society, not an outsider collecting data to take elsewhere. The people I interacted with often explained they maintained their stance and critiques of the EU. As long as there were no funding mechanisms involved, the subjects I followed and spoke to accepted to participate in the research relationship. Critical questions about where the knowledge I build goes and who it benefits remain, urging me to practice ways of making such knowledge available and accessible to the communities I write about here. This research has put me in the conditions of allowing discomfort to open intellectually honest questions based on the respect for my interlocutors, who are not simply "participants" in a knowledge I am an expert in but are knowledgeable themselves in their own experiences and ideas.

Therefore, my relationships with the different groups are contextual: I have not adopted a universally comprehensive model of establishing and practicing research. With Taghmees our relationship in this moment in time was framed as a co-learning experience from different positions. With Al Barakeh Wheat and Farfaheeneh, who carry out less frequent public activities and were available for dialogue in different ways, I joined their spaces and collective activities, rather than including them in the researcher's participatory method.

4 Food Geographies Research as Relational and Embodied Learning Practices: Notes on Methodology

Acknowledging presence and relationship as constitutive to research carries important implications on the theory of knowledge and on methodology. It is not only I who has questions that the "field" shall answer; rather, my co-learners/interlocutors and I navigate situations, each asking and answering from different standpoints.

Food, which we all engage with on a daily basis, can be a particularly interesting research site, linking the matter of food and body/ies interacting with it (their viscerality and living senses), the symbols of meaning-making and knowledges built about and around food, and the socio-economic structures of food production and consumption. Abbots argues against food being approached only as a lens of something else and reclaims food itself as a worthy enough subject of analysis and questioning (Abbots, 2015, pp. 2–3). While this research does look at food as a space of entangled relations and practices, it also acknowledges the methodological opportunities offered by its necessarily intimate relationship with our bodies. Food can represent an "organic" site of knowledge, one where the latter can be exercised as praxis in a Gramscian sense, as it necessarily involves our bodies, transforms them and is transformed by them (Abbots, 2015). In this relationship, the body in its turn acts upon food, others and society, and thinking and knowing happen within our interactions with food, even when unspoken.

In this research—where food becomes a multi-layered site of learning, coming together, performing disruption of hegemonic temporalities and spaces—bodies are always involved in the making of thoughts, which can never be intended as solely intellectual work. This has shaped the ways and times of doing research with my interlocutors. Given the embodied practices of the different groups I engage with, I participated in their activities instead of inviting them into research "participatory methods". This has meant choosing to carry out long, sometimes slow, research and reconfiguring tools like community mapping and Photo-Voice—included in the research project's design—based on the times and sensibilities of subjects and spaces. In this sense, the participatory method was pursued most radically in taking me, the researcher, to the spaces organized by those subjects without introducing my solicitation to doing things together. Methodology has involved ethnographic approaches, but it was the specific pedagogic practices carried out in those spaces that have been most significant ways of learning, contributing to my understanding of things. Walking, eating, cooking, doing, talking within these groups' activities were all meaningful forms of collective learning and knowledge that speak to decolonizing methodologies, food and bodies' corporeality and viscerality research (Hall et al., 2020), and "follow the thing" approaches (Cook, 2004). Qualitative semi-structured interviews—conversations—have also helped reconstruct timelines. Finally, instead of expecting my interlocutors to build a "community map" for my research, I created a personal map: a story of encountering these people and places, and learning and moving with them. The outcome of my research process is intended as my navigation of relationships and meaning-making, and I do not impose it on my co-learners but rather offer it as a contribution to the collective learning processes of a society I am part of.

5 Making Food and Meaning Together: Three Experiences in Amman

Many times, when family members recount memories, the changing extension and expansion of Amman is part of their stories. Remembering visits to aunts and uncles in areas such as *Sweifieh,* or what now-densely populated areas looked like in their childhoods, is often marked by surprise at how distant and difficult to reach those places felt. Amman used to end at ***ad-duwwar ar-rabi'*** (the Fourth Circle), they tell me; all that came after was perceived as far away, physically characterized by expansive wheat fields. This landscape is recalled in my relatives' memories up to the late seventies, when Amman began expanding westward and urbanizing rapidly in the following decades. I approach these memories as "a small door" (Tomba, 2017) that allows the past to seep into the city's present, speaking to and opening questions about it, and the presence of food in it.

5.1 Reconnecting to Food and Land: Questions of Space and Time

Amman's densely trafficked streets dilate travel times and compress the lives of commuters. The city, mainly built for cars, offers little space for walking and is dominated by vehicle rapid proliferation. Drivers and passengers daily spend long hours commuting for work, errands, or leisure. Temporal compression is mirrored spatially: leisure is concentrated in spaces of capitalist consumption and in gentrified, touristified areas of the city— especially in West Amman. Socio-economic divides are starkly spatialized between the latter and East Amman, and within both areas as well. At the same time, open and public spaces of collective encounter and public presence have often been converted into heritage sites (Bustani, 2023a).

This chapter argues that the neoliberal rhythm and space production of West Amman in the past decades has had an impact on the relationships of its inhabitants with food, affecting what they eat, buy, know about it, how they consume it, the ways they exercise political and social choices through it, etc. In this context, the three experiences I engage with are different in their activities, composition, aims and practices, but they all share—in my experience of their work—making social and physical spaces that, even when temporary, possibly open questions concerning the hegemonic neoliberal, aid-dependent and capitalist order of space and time, and perform other rhythms, words and worlds.

While these experiences become within the temporality of the neoliberal city, they share embodying relationships with food that revolve around seasons and seasonality. This re-configures not only their relationship with food but also with land, and therefore with the production of territory. Bringing the question of land into the awareness of people participating in these groups' activities does not happen by romanticizing pastoral life; rather, it implies opening an understanding of how neoliberal development in Amman and beyond has impacted local food systems, stripping people of knowledges and active social and food-land relationships, as well as of the ability to pursue food sovereignty. The organizers of the three groups, in different ways, assert a need to reappropriate an organic relationship with food from within their environments as a means of reweaving what capitalist economies of food have ruptured.

One of the founders of Al Barakeh Wheat spoke about this extensively on a bright early morning in January 2025, when we gathered on land in the Abdoun Corridor area. The *baladi* wheat project had published a call in December 2024—as in previous years—inviting participation in wheat sowing and harvesting throughout 2025. Participation was subject to a financial contribution which the organizers motivated as a means of supporting the costs of the activity itself. Over 120 people—young and older, families, friends and individuals, mainly inhabitants of West Amman—showed up to learn through practice about wheat cultivation. The organizers stood in the middle of the circle of people and explained why they felt it was vital not only to expand local wheat production but also to reclaim space for it inside the city. Their account of what led them to this project, and why it was well-received by people, centred a critique of international financial institutions' conditionalities weakening Jordan's food sovereignty and deepening its food insecurity. Without poetic intent, their emphasis on reclaiming space to cultivate wheat and use it in Jordanian kitchens spoke of material and relational reappropriation. Following the wheat seed from when it is cast in the soil to when it sprouts, grows, is harvested, grinded, packaged, sold and finally used in the kitchen, changes the times and spaces through which participants think of, view and relate to food: it is seasons, rainfall, environmental and climate conditions, social knowledges about how to cook and eat the product, and interdependence, that come forward into the awareness of the eater and consumer, as the organizers noted during the sowing activity. This was reiterated in my conversations with them in June 2025, when we gathered again to harvest the wheat we had sown in January; and was similarly affirmed by one of the two founders during an interview in July 2024, when she spoke to me about how producing *baladi* wheat is no sentimental endeavour for her, but rather reconnects people with the material rhythms of agriculture, land, food and the material resources needed to develop the project free from foreign funding. Indeed, I thought of this during the harvest activity, bent over collecting golden ears of wheat. My body spoke before words, grounding me into the effort of the labour we were carrying out: working the land is hard, a historically collective activity around which social and cultural life developed in Bilad al-Sham. On another level, I also sensed how the temporality of seasons and land, my heartbeat under the sun, and the movement of my and other bodies, all introduced multiple rhythms in that moment, layering the space with different frequencies and stories.

The temporality of seasons is central also in both Taghmees and Farfaheeneh, where gatherings around food align with the calendar of making *mouneh* ("provision", "stored food"), the practice of preserving seasonal vegetables, animal products and herbs through various methods during different times of the year. Jams and marmalades, vegetable pickles, dairy products, molasses and syrups are some of the food products that people have traditionally stored and preserved in Bilad al-Sham. During my time with Taghmees, I had the chance to engage, alongside them, in preparing *mouneh* products (white cheese, *labaneh*, pomegranate and carob molasses, olives and *makdoos*), which they mainly sell at the *Souq nos al-ʿosbou* ("the mid-week market"), co-organized with *Fann wa Chai*, a tea bar and art gallery in Al-Weibdeh. The market, as well as a delicious *mouneh*-based weekly Friday breakfast, takes place involving other local food producers and artisans. For Taghmees, *mouneh* is not only a source of income; it has also been practiced in past years as a learning practice with participants in *tejwal* (Taghmees, 2023) to different areas in the country during specific seasons, offering the possibility of learning about the processes involved in the production of seasonal foods like oranges, pomegranate and cocoa.

Farfaheeneh, a non-permanent space of *lamma* ("gathering") organized by a nutrition researcher who focuses on centring local knowledges, cultures and memories of food in her work, also revolves around seasons. The gatherings are indeed held in the specific times of the year when the land gives fruit, the animals give milk or the produce of the previous year is converted and stored in preparation for the upcoming harvest: pomegranate molasses, goat cheese and olive oil soap are the products that Farfaheeneh focused on between autumn of 2023 and summer of 2024. Along more-than-human and seasonal temporalities, there is also the time of the *lamma*—produced through a coming together of participants. These moments of collective learning produce, in their turn, the re-opening of a "small door" on a past made present by embodying activities of food storing and making that were practiced in almost every household a few generations back, and that have been left behind with the expansion of supermarkets, restaurants, fast-foods and delivery services.

Multiple temporalities are re-activated in the work of these three experiences, based on relationalities that produce spaces of re-connection between humans, non-humans and more-than-humans that question the fragmentation of time, self and community in the rapid rhythm and development of the neoliberal city.

5.2 *Detaching from Funding: Redefining "Abundance" in Land-Rooted, Socially Embedded Ways of "Being-in-Common"*

"We were taught in school that Jordan is a resource-scarce and poor country", one of Taghmees's founders tells me as we drive to the village of Judayta, in the north, to meet a local producer of carob molasses. I had learnt the same sentence in primary school.

The founders of Taghmees have often shared the story of how the idea of starting this learning experience and space emerged. After years in local and community development NGOs, they grew radically critical of development and the dependencies it produces in Jordanian society and globally. Leaving that world and moving toward the unknown of imagining and doing otherwise (or, as they always point out: "going back to what we already knew and was silenced and taken away from us", underlining the need to detach from perceiving oneself as an *alter* to a hegemonic self) has been described by them as a turbulent process of re-learning how to be in the present, both temporal and ontological, by rejecting the linearity and outcome-based mindset of project-planning.

From the scarcity taught in school and institutional *ta'leem* (education), they wanted to move toward *ta'allum* (learning) through *taghmees* (immersion). The epistemic, material and ontological shift required to do so is well expressed in a popular proverb they have adopted: "*Al-jood bil mawjood*", meaning "generosity is present in what we have"—a saying that underscores giving as a foundational act and part of prosperity. This radical outlook on the relationship with "resources" characterizes the entire practice and space that Taghmees has been exercising in the past twelve years, translating into a total rejection of international development funding and its dynamics and conditionalities. This is where I have personally found one of the most interesting elements of their work: the very material tension and urgency that arises when rejection is practiced. Closing doors of interaction with that kind of funding is not simply an ideological positioning, it has material implications and effects on the activities and spaces that Taghmees was able to enact—or not— in time. In a moment when the opposition to funders like USAID, the EU and governments expressing support for the Zionist aggression on Gaza, expanded in Jordanian society, Taghmees shared their critique of these institutions and of development as a material and epistemic worldview (Taghmees, 2024). It is within this critique— stemming both from their critical studies of development and a decade-long experience—however, that Taghmees needed to imagine and do something different that could allow it to keep on offering spaces of collective learning inside and outside the city. An answer was found in making space for a mid-week market involving other local small producers of food, organic products and artisanal work in Jabal Al-Weibdeh. In my experience of being in this physical space, this was a place where economic transactions were articulated in the language of collaboration instead of competition, and where interactions were often animated by moments of collective singing, dancing and storytelling. Practicing the market in this way resonates with a post-capitalist sensibility in that there is an attempt of liberating economic transactions from capitalist imperatives, while still being aware of entanglement in the capitalist economic system. The market may be perceived as a place that opens a crack in the present, a space for doing differently in the matter of daily life and experience. It does not enact a full liberation from capitalist economy and gentrified urban space (nor does it aim to produce a total transformation at once). However, its materiality and relationality function as a performance of disruption of capitalist market linearity and spatiality, offering questions on food systems, markets and knowledges in the present of the city and the world.

This *disruptive performance* also appears in Al Barakeh Wheat and Farfaheeneh's spaces of food production and selling in the first case, and collective learning in the second. Both groups articulate a critique of international funding and its dependencies, and they materially enact alternatives. Gathering people to follow *baladi* wheat from field to flour, to supermarket shelves, bakeries and restaurants, or re-learning practices of food conservation, performs a disruption of the neoliberal production of the city's space and food. Cultivating wheat where it was once harvested—making food sprout instead of shopping malls—is suggestively represented in a photo taken during one of the past collective harvest activities organized by Al Barakeh Wheat in Amman. In the picture, a land of golden wheat is harvested by participants standing out against the huge buildings of City Mall in the background (Bustani, 2023a, p. 13). I saw a similar contrast in this year's harvest, as the land we gathered on stood by a highway and the arches of a McDonald's waved in the background of a picture I took. On the other hand, re-learning socially embedded food storing and production techniques, in Farfaheeneh's case, performs a disruption of the predominantly consumerist relationship with food in the city, reintroducing the temporality of making the products and conserving them, as well as the organizer's critique of mass-produced foods, which she articulates with participants during the gathering.

In different ways and spaces, the three experiences resonate together around the critique of development's function, and in the practice and embodiment of ways of detaching from the matter and worldview produced by it. In doing so, they go back to the roots of food and social fabric, where they attempt to reappropriate and reactivate relationships and knowledges that redefine "abundance" and "scarcity". I speak of "performing disruption" to point to how these practices embody interruptions of capitalist matter and temporality—albeit in different ways and permanencies—while I also recognize how such disruptions remain on a performative/narrative, rather than structural, level. Nonetheless, these performances of a pluriverse of meanings and relationships with food, land and community articulate and embody critical questions concerning global liberal economies, neoliberal space production and their impact on food sovereignty and epistemic and material dependency in Jordan.

5.3 *Local Pedagogies and Epistemologies of Food: Being in a Convivial Pluriverse*

The learning practices encountered in these three spaces are, perhaps, what resonates together most clearly and immediately, leading me to look at them in constellation, a method derived from Walter Benjamin's critical perspective on modern history, and meant as a configuration of experiences that speaks "a time of the now" that questions the processes historically leading to the present moment (Benjamin, 1940, p. 12).

While the three experiences differ in aims and rhythms, what I have consistently found while participating in their activities is the effort to embody learning spaces that function in two directions at once: critically questioning prevailing narratives about food, the city and the inhabitants' relationship with food and land; and practicing meaning-making and knowledge-building rooted into social relations and embodied practices. This reappropriation of a living and transformative relationship with food, community and land opens radical questions about ways of learning and knowing, as meaning emerges through relations and practices not always recognized or deemed as legitimate pedagogical, epistemological or research methodologies by academic disciplines.

In Al Barakeh Wheat's case, the collective harvest activities carried out inside the city are a learning space where the body-land relationship is practiced against the grain of the neoliberal matter of the city. *Barakah* ('blessing') is not an abstract concept but an embodied, material philosophy of interdependence between humans and with non-humans, a holistic way of working the land and conceiving life. One of the project's founders explained this while describing *zakat al qamh* ('the wheat charity')—the Islamic obligation of giving a portion of the wheat harvest as charity—which farmers often practice in a more-than-human configuration: by throwing a portion of seeds in the land for other beings (birds, ants, insects, etc.) so they are fed from this *rizq* (provision) within human sustenance through agriculture. Moreover, on the occasion of the 2025 harvest activity, the organizers declared the *zakat al qamh* (amounting to 10% of the proceedings) would be dedicated to supporting families facing economic challenges. The philosophy of *barakah*, in this sense, and its sharing through collective sowing and harvesting inside the city, become practices of knowledge and learning that counter the progressive, extractive and accumulation-based epistemology and materiality of capitalist food production and consumption. In Farfaheeneh's case, the gatherings (*lamma*) allow for a reconstruction of a relationship with food and seasons: anchoring learning to seasonality connects the participants' bodies with the times of the land and food-making. Again, the temporality of storing disrupts the time of buying and immediately consuming food, centring learning around the slow pace of chemical and social processes of making food.

Attention to social stories and fabric is central also in Taghmees's idea of "immersion". Over twelve years, the practice of sitting around shared food and eating while conversing and telling stories around an open question has enabled the group and its *ahl* ('kin') to learn across diverse spaces animated by *jaw taghmees* ('the atmosphere of *taghmees*'), made present by a banner permanently hung in their space. The banner speaks invitations to follow during gatherings: no judging, no interrupting, no advice, being present, listening, no repressing, respecting each other, speaking about one's own self. These engagement rules guide the learning experiences and the participants: in *tejwal* (walks), *mujaawarah* (learning through neighbouring), *taghmeeseh* and "social mixes", learning emerges through the sharing of personal and collective stories instead of debating on a theoretical level alone. This is meant to root learning into practice and to recover words that the plastic language of modern education institutions and development make invisible in societies made materially and epistemically dependent.

In all such cases, food becomes the container and space of relationality— with body, environment, society and community— and serves as a learning space not only for participants, but also, consistently, for the organizers themselves, who shared with me their learning processes, and continue to reflect on and share them on their platforms.

6 Conclusion

This chapter has attempted to respond to a thesis that neoliberal urbanization, international development dynamics and liberalized economies inform the practices and relationships of Amman's inhabitants with food, as well as the foodscapes of West Amman. The expansion of the city in the past decades has affected its environment as well as the lifestyle of Ammanis. In this context, the three experiences explored in this chapter recall and embody local foodways, in different ways, that disrupt the neoliberal city's hegemonic temporality and spatiality through practices of eating, making food and sharing, selling, storing and knowing it. In this sense, they offer a "pluriversal" image of times, spaces and knowledges in Amman that counters the predominant narrative and materiality of the city.

Tensions and open questions remain: more research needs to be carried out on food in connection to the socio-economic differences and divides characterizing West Amman and the city at large. Participants in the experiences I have followed do not represent the entirety of the city's inhabitants. Moreover, research on food spaces and dynamics in East Amman would bring to light further critical questions on several levels, deepening the understanding of the city's development. What emerges in this research, I argue, is the highly insightful, socially charged and epistemologically stimulating character of food in the context of Amman: how it speaks about social, political, spatial and temporal relations. Within spaces, matters and symbols of food, we can read both the development of capitalist economic configurations and worldviews as well as practices and discourses countering them, which focus on rebuilding organic relationships with land, social fabric and stories. The latter reclaim material, ontological and epistemic worlds shadowed by the sparkling discourse of progress and development, and possibly disrupt spaces and times of the unilinear capitalist global order by practicing aspirations for food justice, abundance and convivial post-humanist interdependence.

References

Abbots, E. (2015). *The Agency of Eating. Mediation, food and the body.* Bloomsbury Academic.

Abdeljawad, N., & Nagy, I. (2021). Urban environmental challenges and management facing Amman growing City. *Review of International Geographical Education (RIGEO), 11*(5), 2991–3010. https://doi.org/10.48047/rigeo.11.05.192

Abu al Haija, A., & Potter, R. B. (2013). Greater Amman: Metropolitan growth and scenarios for sustainable urban development. In C. Gambardella (Ed.), *Heritage, architecture, Landesign: Focus on conservation, regeneration, innovation, Le vie Dei Mercanti: XI forum Internationale di Studi.* La Scuola di Pitagora Editrice.

Abu Aqiqah, K. M., & Al-rfou', A. N. (2010). The role of tourism sector on economic development in Jordan during the period (1990–2008). *European Journal of Economics, Finance and Administrative Sciences, 18*, 173–170.

Al Akayleh, F. (2013). Agricultural development in the age of trade liberalisation: What did we really get? A case st:Udy of Jordan. *International Journal of Economics and Business Research, 5*, 3286–3301. https://doi.org/10.1504/IJEBR.2013.052484

Al-Assaf, G. (2025). Foreign aid: Challenges to local economy of civil Society of Jordan. In I. Natil (Ed.), *Civil society and local. Ownership in the global south* (pp. 94–108). Routledge.

Albaik, M. (2023). The social impact of gentrification in the Jabal Al-Weibdeh neighborhood in Jordan. *ISVS e-journal, 10*(5).

Al-Jazeera. (2023, November 16). *Jordan says it won't sign energy and water exchange deal with Israel.* Al Jazeera. Retrieved July 16, 2025, from https://www.aljazeera.com/news/2023/11/16/jordan-says-it-will-not-sign-energy-and-water-exchange-deal-with-israel

Al-Jazeera. (2025a, November 25). *US-backed GHF 'aid mission' in Gaza ends—a timeline of violence.* Al Jazeera. Retrieved October 20, 2025, from https://www.aljazeera.com/news/2025/11/25/us-backed-ghf-aid-mission-in-gaza-ends-a-timeline-of-violence

Al-Jazeera. (2025b, May 20). *What is the Gaza Humanitarian Foundation, and why has it been criticised?* Retrieved July 16, 2025, from https://www.aljazeera.com/news/2025/5/20/what-is-the-gaza-humanitarian-foundation-and-why-has-it-been-criticised

Al-Johari, E. E., & Sharaf, F. M. (2024). A land suitability model of urban agriculture: The case of Amman City in Jordan. *Universal Journal of Agricultural Research, 12*(2), 211–224. https://doi.org/10.13189/ujar.2024.120201

Beauregard, R. A., & Marpillero-Colomina, A. (2010). More than a master plan: Amman 2025. *Cities, 28*(1), 62–69. https://doi.org/10.1016/j.cities.2010.09.002

Benjamin, W. (2005; [1940]). *On the concept of history*. Gesammelten Schriften I:2. Suhrkamp Verlag. Frankfurt am Main, 1974.

BNC (2025, November 5). "New Victory: Carrefour closes all branches in Jordan due to BDS pressure", BDS. Retrieved July 16, 2025, from available at: https://bdsmovement.net/Carrefour-Closes-All-Branches-In-Jordan-BDS-Pressure (16/07/2025).

Brighenti, A. M., & Kärrholm, M. (2018). Beyond rhythmanalysis: Towards a territoriology of rhythms and melodies in everyday spatial activities. *City, Territory and Architecture, 5*(5), 4. https://doi.org/10.1186/s40410-018-0080-x

Bustani, H. (2023a). *Waking up to my destroyed city*. The Arab Institute for Research & Publishing.

Bustani, H. (2023b). *(Dys)functional polities: The limits of politics in the postcolonial Arab region (PhD thesis)*. University of Westminster Humanities. https://doi.org/10.34737/w67yx

Bustani, H. (Ed.). (2024). *An taktub al-Madina: Amman min al-masafa sifr (trad. Writing the City: Amman from zero distance)*. The Arab Institute for Research & Publishing.

Bustani, H. (2025). Preparing for revolutionary times? Chronic crisis of authority and constructive subversion in contemporary Jordan. *Interventions*, 1–19. https://doi.org/10.1080/1369801X.2025.2504923

Cook, I. (2004). Follow the thing: Papaya. *Antipode, 36*, 642–664. https://doi.org/10.1111/j.1467-8330.2004.00441.x

Counihan, C., Van Esterik, P. V., & Julier, A. P. (2019). *Food and culture. A reader*. Routledge.

Daher, R. (2008). Amman: Disguised genealogy and recent urban restructuring and neoliberal threats. In Y. Elsheshtawy (Ed.), *The evolving Arab City. Tradition, modernity and urban development* (pp. 37–68). Routledge.

Daher, R. (2011). Prelude. Understanding cultural change and urban transformations: Qualifying Amman: The City of many hats. In *Villes, Pratiques Urbaines Et Construction Nationale En Jordanie*. Presses de l'Ifpo. https://doi.org/10.4000/books.ifpo.1725

Daher, R. (2013). "Neoliberal urban transformations in the arab city Meta-narratives, urban disparities and the emergence of consumerist utopias and geographies of inequalities in Amman". Urban Environment. Volume 7: *Villes arabes, villes durables? Enjeux, circulations et mise à l'épreuve de nouvelles politiques urbaines.* 99–115. doi:10.7202/1027729arCopied.

Doan, P. L. (2006). Tourism planning and regional instability. The consequences of rapid expansion of the tourism sector in Jordan. *International Development Planning Review, 28*(3), 311–332. https://doi.org/10.3828/idpr.28.3

Escobar, A. (1995). *Encountering development. The making and unmaking of the third world.* Princeton University Press.

Escobar, A. (2018). *Designs for the Pluriverse: Radical interdependence, autonomy, and the making of worlds.* Duke University Press.

Fasheh, M. J. (2021). *Ihtilal Wa 'awda. Rubbama tabhath bayn al-ghusun 'an ma yumkin al-'uthur 'alayh faqat fil juthur ("occupation and return: Perhaps you are searching between the branches for what is found in the roots").* Al-Dar Al-Ahlia Bookstore.

Fisher, M. (2009). *Capitalist realism: Is there no alternative?* Zero Books.

Gaul, A., Auman Pitts, G. (2021). "Introduction: Making Levantine cuisine", 1–22, in Gaul, Auman Pitts, Valosik (Eds.) Making Levantine cuisine. Modern Foodways of the eastern Mediterranean. University of Texas Press.

Gharaibeh, A. A., Al-Zu'bu, E. M., & Abuhassa, L. B. (2019). Amman (City of waters); policy, land use, and character changes. *Land, 8*(12), 195. https://doi.org/10.3390/land8120195

Gibson-Graham, J. K. (2006). *A Postcapitalist politics.* University of Minnesota.

Gramsci, A. (1971). *Selections from the prison notebooks (edited and translated by Quentin Hoare and Geoffrey Nowell Smith).* Lawrence & Wishart.

Hall, S. M., Pottinger, L., Blake, M., Mills, S., Reynolds, C., & Wrieden, W. (2020). Food for thought? Material methods for exploring food and cooking. In *Mundane methods.* Manchester University Press. https://doi.org/10.7765/9781526152732.00012

Hammelman, C., Levkoe, C. Z., & Reynolds, K. (2024). *Radical food geographies: Power, knowledge and resistance.* Bristol University Press. https://doi.org/10.2307/jj.9692676

Hopma, J. (2017). *"No one can solve this problem but the modern prince". Counter-hegemonic discourses of Food Security & Food Sovereignty in Jordan: A Gramscian analysis.* (Doctoral Dissertation),. University of Aberystwyth.

Hourani, N. (2014). Assembling structure: Neoliberal Amman in historical perspective. *Urban Anthropology and Studies of Cultural Systems and World Economic Development, 45*(1/2) Special issue: State and Nation in the Arab Middle East, 1–62.

Human Rights Council (2024). *Anatomy of a Genocide. Report of the Special Rapporteur on the situation of human rights in the Palestinian territories occupied since 1967, Francesca Albanese.* Fifty-fifth session. 26 February–5 April 2024. Agenda item 7. Human Rights situation in Palestine and other occupied Arab territories.

Illich, I. (1973). *Tools for conviviality*. Harper & Row.
Jordan News Agency (2025, March 24). *GAM approves urban planning regulations for 2025*. Jordan News Agency. https://www.petra.gov.jo/Include/InnerPage.jsp?ID=69732&lang=en&name=en_news
Kably, A. (2015). *Learning From Self-Planned Communities*. TUCAC15 Paper–27.
Lefebvre, H. (2004). *Rhythmanalysis: Space, time and everyday life (or. Title: Elements de rhythmanalyse, 1992. Paris: Edition Syllepse)*. Continuum.
Mbah, R. E., Hardgrave, C. M., Mbah, D. E., Nutt, A., & Russell, J. G. (2025). The impact of USAID budget cuts on global development initiatives: A review of challenges, responses, and implications. *Advances in Social Sciences Research Journal, 12*(04), 219–232. https://doi.org/10.14738/assrj.1204.18681
Ministry of Environment. (2017). *National Green Growth Plan for Jordan*. Hashemite Kingdom of Jordan.
Newton, L. (2022). Epistemic anxiety and epistemic risk. *Synthese, 200*, 324. https://doi.org/10.1007/s11229-022-03788-7
Page, J. (2013). *Rainbow street: Heritage, cosmopolitanism, and urban space in Amman, Jordan. The inquiry. The journal of the Mellon mays undergraduate fellowship, the Center for the Humanities and the College of Arts and Sciences* (pp. 80–108). Washington University in St. Louis.
Schuetze, B., & Hussein, H. (2023). The geopolitical economy of an undermined energy transition: The case of Jordan. *Energy Policy, 180, 113655.*, ISSN 0301-4215(2023). https://doi.org/10.1016/j.enpol.2023.113655
Schwedler, J. (2010). Amman cosmopolitan: Spaces and practices of aspiration and consumption. *Comparative Studies of South Asia*, Africa and the Middle East (2010), *30*(3), 547–562. https://doi.org/10.1215/1089201X-2010-033
Shami, S. (1996). Re-searching the City: Urban space and its complexities. In J. Hannoyer & S. Shami (Eds.), *Amman*. Presses de l'Ifpo. https://doi.org/10.4000/books.ifpo.8224
Shami, S. (2007). Amman is not a City: Middle eastern cities in question. In A. Çınar & T. Bender (Eds.), *Urban Imaginaries* (pp. 208–235). Locating the Modern City.
Taghmees. (2023). *Tejwal*. Taghmees. Retrieved July 16, 2025, from https://taghmees.org/category/tejwal-%d8%aa%d8%ac%d9%88%d8%a7%d9%84/
Taghmees. (2024, October 16). *Masyadat at-tamweel al-ajnabi* ("The trap of foreign funding"). Instagam post. Retrieved July 16, 2025, from https://www.instagram.com/p/DBMHSAftQtD/?img_index=1
Tammey For Youth. 2024. *Tammey For Youth Development*. Retrieved July 16, 2025, from https://www.instagram.com/tammey_4_youth/
Tomba, M. (2017). *Attraverso la piccola porta. Quattro studi su Walter Benjamin*. Mimesis Edizioni.

Tuhiwai Smith, L. (2012; 1999). *Decolonizing Methodologies. Research and indigenous Peoples*. Zed Books.

World Food Programme (2025). *WFP runs out of food stocks in Gaza as border crossings remain closed*. WFP. https://www.wfp.org/news/wfp-runs-out-food-stocks-gaza-border-crossings-remain-closed

Yousef, M. (2024). Tourism in Jordan's economy: Growth, challenges, and future prospects. *International Multilingual Academic Journal, 9*(5).

CHAPTER 4

Urban Marine Foodways: Walking and Sensing Artisanal Fishing along Tangier's Waterfront

Ginevra Montefusco

Abstract This chapter aims to trace the pressures on Tangier's marine foodways by adopting the mobile, visual, and participatory method of urban itineraries. Specifically, it focuses on the urban waterfront of Tangier as a place of friction where various ecological and political processes that hinder the practice of urban artisanal fishing converge. The chapter frames artisanal fishing in Tangier through the lens of Radical Food Geographies and urban political ecology, defining it as a relational practice which connects dwellers with urban natures and marine ecologies. Secondly, the chapter unfolds urban itineraries as a participative and embodied method for investigating urban food places and stories, advancing methodological discussion around the field of urban food geographies. By exploring three itineraries co-designed with research participants in Tangier through mobile interviews (walking or driving),

G. Montefusco (✉)
Department of Historical and Geographical Sciences and the Ancient World (DiSSGeA), University of Padua, Padua, Italy
e-mail: ginevra.montefusco@phd.unipd.it

P. Minoia (ed.), *Foodways As Sites of Knowledge and Transformation*, Emplacing Food, https://doi.org/10.1007/978-3-032-28126-5_4

the chapter will highlight the role of urban development and more-than-human connections in shaping artisanal fishing practices. Each itinerary will unfold the story of core urban places that are related to local foodways and their transformation, focusing on urbanisation dynamics, ecological erosion, and community fragmentation in coastal neighbourhoods. This approach aims to visualise the entanglements animating the city of Tangier, its waterfront, and artisanal fishing practices. At the same time, these itineraries will trace the ongoing processes that extend beyond the confines of the waterfront, discussing impacts of urbanisation dynamics, such as ecological erosion, and community fragmentation in coastal neighbourhoods. This paper starts from fishing places and practices— places of work as well as leisure, practices of subsistence as well as conviviality— understood as knots where the different dimensions of foodways are intertwined and transformed. Through participatory geographical observation mediated by movement and photography, a mobile and trans-scalar perspective on foodways is adopted: from political and ecological processes to everyday experiences and grassroots relations, from wider transitions to specific frictions, and back.

Keywords Artisanal fishing · Socioecological relations · Mobile methods · Food places · Urban natures

1 The Rods, the Grass, and the Fence: A Sensory Introduction to Tangier's Waterfront

At nine in the morning, the tide is low, and the rocks in front of Tangier's Corniche start emerging. It is the ideal time for fishermen to climb over the wall that separates the high-traffic road and the sea, and advance towards the first row of rocks with rods on their backs. On the other side, a vast lawn extends along the boulevard. The first municipal workers arrive to water the Kikuyu grass, amidst compositions of aloes, palms, and exotic flowers. Along the *corniche,* three projects at the heart of Tangier's urban development are unfolding. The new artisanal fishing port is invisible from the concrete promenade, hidden behind high fences as a proper border zone. Following this path, the new Lalla Abla Mosque, as well as the new Marina Yacht Club, inaugurated in 2018, will appear. Interspersed with these major works are several construction sites for resorts and shopping malls promising, in adjacent billboards, "uncompromising luxury". From the ramparts of the kasbah, the highest part of the medina, it is possible to grasp with one sight these traces which compose, as a mosaic, the new narrative that Tangier makes of itself—a modern, attractive city where "a new story is about to begin", as other banners state. Seen from these walls, namely the historical fishermen's neighbourhood called Bab el Bahr, this "new story" can be perceived as distant and, to some, unfamiliar. Through its architecture and visual narrative, the waterfront exhibits, as in other redeveloped Mediterranean port cities analysed in Duruz et al. (2022), "a certain 'sameness' generating a 'sense of déjà vu'" (p.574). The view, which frames both the sea and the new façade, can evoke a range of emotions in Tangier's dwellers as well as visitors, including feelings of humiliation and anger. The Strait of Gibraltar separates Tangier from Tarifa by 17 nautical miles. The new port, following the structural transformation of the fishing sector in the kingdom, is a highly securitised site, suggesting the tightening of the border between Morocco (and, by extension, the continent) and Europe. Artisanal fishing, and consequently the port, embody a history of informal economies, clandestine migration, as well as political tension between the city (and its region) and the central government, all of which marked Tangier's life and identity for several decades after Morocco's independence from European colonialism in 1956. The symbolic relationship between the work of fishermen and the region's uprisings has cyclical returns; the latest is the case of Mouhcine Fikri, namely an informal fish seller who died in 2016 in nearby coastal Al-Hoceima while trying to retrieve his cargo of fish, which was confiscated and thrown into a rubbish compactor by a policeman (Vacchiano, 2022). This event, in the wake of a complex history of the region's dissidence to the central government and Spanish colonialism, triggered the Popular Movement of the Rif, namely a collective response to the Fikri scandal. The movement's demands also included integration into national development plans, increased job opportunities, and improved infrastructure to counteract unemployment. In the ranks of these demonstrations, signs read: *al-hogra taqtal*, humiliation kills (Vacchiano, 2022).

Walking along this porous threshold allows us to grasp ecological elements, fishing practices, and spatialised policies from a sensory and embodied perspective. The sharp smell of the fish restaurant at the edge of the port, the embodied discomfort produced by the surveillance technologies enclosing the area, and the noise of cars now moving at higher speed along this new road, all serve as crucial clues for understanding the city's development trajectories and their impact on interstitial practices, such as small-scale fishing. Indeed, Tangier's waterfront offers a telling reflection of the kingdom's development model, which can be traced in programs like Ville sans bidonvilles (2003) and Tanger Métropole (2013–2017). These regeneration plans encompass several political agendas, including the tightening of borders, the enhancement of the economic asset of fishing, and the regeneration of urban space vis-à-vis tourism and major events, such as the African Cup of Nations (2025) and the FIFA World Cup (2030). This space, so densely political in a material and symbolic way, has been analysed as the outcome of structural historical-political processes that make Tangier a global city, interconnected with the rest of the country, as well as with the Mediterranean (Wippel, 2022). In the meshes of the neoliberal narrative of development and modernisation materialised along the waterfront, artisanal fishing represents a space of friction. Even though less accessible than before the new development project, numerous fishing rods are visible along the corniche, somewhat hidden between the breakwaters.

To delve into urban marine foodways and their transformation, urban food geographies are called upon to implement embodied and transformative methods able to focus on the places where food practices are rooted and connect them in wider processes of change. The chapter proposes a perspective on urban foodways with a visual, mobile, and narrative approach. Starting from its food places, the city will be explored through a trans-scalar lens: from structural processes to everyday experiences, to wider transitions to specific frictions, and back. This essay proposes an introduction to Tangier's foodways, exploring three itineraries co-designed with seven research participants—fishers, activists, and practitioners—through mobile interviews (walking or driving). Each itinerary will unfold the story of core urban sites that are related to foodways and their transformation. The first itinerary will focus on urban development, touching on the main areas of change in the city, its drivers, and the impact on fishing practices of self-sufficiency and leisure. The second will tackle the ecological and more-than-human dimensions of these changes, emphasising the relationship between the sea, urban green and waterways as experienced by fishermen. The third will explore the transformations of Bab el Bahr, a neighbourhood deeply impacted by the waterfront regeneration, where the nuances of foodways related to spirituality, belonging, and commoning are more tangible. Through these itineraries, artisanal fishing will be unfolded as a relational practice which connects concerns around marine ecologies and urban development. These walks will highlight the entanglement between land and sea, as well as the assemblage of places and actors which co-produce local foodways. The analysis of these itineraries draws on urban political ecology and the framework of Radical Food Geographies, namely a theory-praxis which intersects spatial and food injustices while tracing socioecological relations embedded in urban food practices.

Before delving into addressing the specific dimension of Tangier's marine foodways—namely, how fish are caught, purchased, consumed, and through which exchange circuits they move across the city and outside of it—this paper starts from fishery-related places and the urban nets that connect them. To make the connections and interdependencies among foodways, urban natures, and their spatial dynamics more tangible, the following itineraries will touch both on places ordinarily associated with fishing—such as the port, the waterfront, and the fishers' neighbourhood—and on the most overlooked and unexpected—namely peri-urban industrial areas and wetlands. As will emerge from this participative and mobile observation, tracing the ongoing spatial shifts can offer a more nuanced understanding of everyday experiences of small-scale urban fishers and the limits they negotiate across work and leisure activities. Therefore, this essay seeks to find a multilayered answer to a complex question: which spatial and ecological processes are impacting Tangier's marine foodways?

2 Urban Waterfronts as Socionatures: Relational Food Practices

Urban waterfronts are complex assemblages where the interweaving of different agents and processes is tangible, from the highly visible to the routinely overlooked. From a geopolitical perspective, they are studied as borders, as well as in an economic and planning key, as sites of investment and infrastructural development. Socioecological approaches to waterfronts, focusing on the intersection of human and non-human ecologies and societies, offer a generative lens to trace how the space between the sea and the city is ordinarily lived and negotiated between neighbouring multispecies communities (Barchetta, 2019; Bunce & Desfor, 2007). In this vein, the literature on socionature examines how nature and society are materially and discursively co-produced through historically situated processes (Alkon, 2013; Moragues-Faus & Marsden, 2017). Although represented as ontologically separate and antagonistic, cities and natures are mutually shaped through metabolic processes, namely by capital flows, accumulation, and extraction dynamics (Swyngedouw, 1996). Shifting the focus towards a more embodied and ordinary understanding of urban metabolism, the so-called second generation of Urban Political Ecology enhances a relational and more-than-human interpretation of the city as a socionature. In this perspective, the co-production process underlying urban natures can be traced in marginal socio–ecological relationships connecting natures, inhabitants, and broader processes of urban planning and development (Grove, 2009; Heynen, 2018; Tzaninis et al., 2020). The ecological and vernacular knowledges of urban dwellers thus become key to understanding how urban natures are transformed or sustained in everyday practices in ways that exceed hegemonic and human-centric representations of urban space and lifestyles (Barua & Sinha, 2017; Wilson, 2024).

Food, though often overlooked by urban political ecology, is the vital synthesis of the co-production process that underlies socionatures in material and cultural ways. Urban oceans, for instance, have long shaped the food habits of coastal inhabitants. In turn, the spaces connecting oceans to land have been historically designed around fishing needs through infrastructures such as ports and harbours. Beyond this material and physical co-production, coastal cities are animated by rituals and practices developed by artisanal fishers over centuries, expressing a visceral and socioecological relationship beyond subsistence or profit logics. Food thus emerges as a crucial entry point to unfold socioecological relations, starting from ordinary food practices that link urban and ocean ecologies. Relationality, at the heart of the most recent debates around food geographies, is well expressed by the anthropological concept of foodways, defined as the set of knowledge and beliefs surrounding the production, distribution, and consumption of food in a cultural context (Counihan, 1999). In other words, food is not just produced and consumed as a commodity but is also imbued with collective and subjective meanings. These meanings are generated, cultivated, and transformed through practices, such as artisanal fishing, which are constantly negotiated with ecological shifts—both ordinary ones, such as changing seasons and tides, and long-term processes, including ocean warming and marine species extinction. Socioecological change, and the relationships it produces, are thus trans-scalar, encompassing different dynamics and actors globally, but have specific spatial and cultural configurations that can be followed through the lens of food. To unfold food as a situated practice, recent food geographies invite us to begin with places where foodways are rooted, attending to how food, ecologies, and place are co-produced (Goodman, 2016; Montefusco, 2026). Pivotal is Brooks and Hubbard's (2022) reflection around the active role played by oysters in changing the material, sensory and affective qualities of place, enhancing gentrification processes in coastal cities, while also being at the centre of local practices and events. Rodenbiker (2023) unpacks the classed, gendered, and racialised relationships that shape the co-production of oceans and cities through the lens of food, bringing the case of the Totoba fish maw market in New York and Hong Kong. In this perspective, small-scale fishing as a relational practice with urban natures is a promising thread to explore. Through their work on urban natures and informal food practices in Seattle, Poe et al. (2014) conceptualise urban foraging as "relational ecologies of belonging". This definition highlights how food practices give meaning to place, embodying knowledges often marginalised by urban planning (Barua, 2014). Taking fishing as an example, food practices entail deep relations with oceans through active learning-by-doing, and with places where these practices are enacted, as they generate a specific sense of belonging to the city. Relational ecologies of belonging add an important layer to foodways, foregrounding how key urban places—particularly at the interface with urban natures—embody meanings, rituality, temporalities, and belongings that exceed food production. In sum, these places are key for sustaining the relationship of subsistence and belonging to the ocean expressed by artisanal fishers and foragers, making ecological erosion and limited access to these places an issue of both spatial and food injustice. Through the lens of urban political ecology and relational food geographies, the analysis of emplaced marine foodways can interweave spatial injustice (i.e. the dynamics that make these spaces less accessible or ecologically damaged) and food injustice (namely, the constraints on self-production and consumption of food in accordance with local, artisanal, place-based foodways). Radical food geographies propose to bridge these two dimensions (Hammelman et al., 2024). This scholarship, sensitive to spatialisation of power dynamics, can support an analysis of political and ecological processes that generate injustice, and their impact on the emplaced relations connecting people, food, and urban natures. Accordingly, the impact of urban development on food commons, place-based solidarity networks, and collective infrastructures is questioned. Building on this conceptual framework, the following section outlines a sensuous and participative methodology for geographical observation, adopted to trace spatial and ecological changes that jeopardise artisanal marine foodways.

3 Walking and Filming: Mobile Methods for Tracing Food Places and Entanglements

Beginning in the 1990s, critical debates around ethnographic methods in geography have been giving generous insights into the embodied aspects of fieldwork, especially feminist, post-structuralist, non-representational, and recent decolonial turns. When it comes to urban contexts, exploring food and place entanglements involves sensuous aspects of place-making, such as taste and smell (Lee & Ingold, 2006). Walking as a method, both collectively and individually, can elicit "a multilayered interaction with space that exceeds the visual realm and extends to the shifting parameters of consciousness, memory, and experience" (Gandy, 2024, p. 1386). This debate aligns with open questions in relational food geographies: how can we explore foodways—namely, knowledges, practices, and relationships that connect the ecological and the social—from a spatial and trans-scalar lens?

Walking is not new as a practice for urban exploration, nor as a geographical method (Merriman, 2014). The urban citizen, the tourist, and the flâneur are key (mainly masculine) figures that emerged from the rise of cities and are related to the practice of walking, embodying freedom of movement and consumption, as well as the paradigm of progress (Cresswell, 2006; Pierucci, 2025). Geography, on the other hand, has developed different methodologies around movement, highlighting how mobile methods can facilitate embodied and lived experience of space—such as transect walks, follow the thing, and interviews on the move. Mobile interviews offer a privileged entry point into participants' everyday places, leaving the power to shape the literal and metaphorical direction of the interview to research participants (Finlay & Bowman, 2017; Holton & Riley, 2014). The new mobility paradigm (Sheller & Urry, 2006), enhanced by feminist and post-representational theories, conceptualises walking as a performative and embodied research practice (Sheller, 2018). Critical debates around walking highlight how an ordinary, mundane, and often defined as non-scientific practice, can foster political discussion about access to urban natures (Mason et al., 2023).

In this study, I apply this perspective to key places where artisanal fishing practices are embedded, adopting mobile and visual methods. Streule (2020), in her mobile toolkit for urban explorations, proposes to work with cities across multiple scales—from the micro-scale of entrevistas en movimiento (mobile interviews) to the metropolitan scale of recorridos explorativos (guided tours across the extended urban space). A geographical observation attending urban shifts and foodway transformation benefits from these juxtapositions and extensions, playing with scale across the city. This zoom-in-and-out process echoes Radical Food Geographies' call for methods that can deepen the relational understanding of emplaced food practices, such as artisanal fishing, while also tracing wider dynamics of spatial and food injustice. On the other hand, the support of visual tools to this mobile approach, which Barron (2021) defines as photo go-alongs, seeks to archive Tangier's spatial transformations as an urgency arising from its complex temporality. Photographing and filming shifting places while walking with participants goes beyond documenting urban change; it facilitates discussions with participants about their lived experiences of these transformations and their implications for everyday food practices. Furthermore, attentive observation through photography allows the researcher to attune to the diverse rhythms of subjects inhabiting and crossing these places (Gandy, 2024). Following post-representational approaches, the narrative emerging from these photographic moments reflects a convergence of participants' perspectives, the researcher's engagement, and the different subjects they encounter while crossing different places (Lorimer, 2010).

The outcome of these mixed methodologies—namely, walking interviews and photo go-alongs—is presented in this chapter through three itineraries. These itineraries are part of an ongoing mapping practice that seeks to trace broader processes of urban change affecting foodways from an underrepresented perspective. Following urban assemblage theory, mapping is a process rather than an output or a representation of pre-existing truths. To make this process grounded and participative, geographical research can benefit from "slow" and sensuous methods as walking interviews and photo go-alongs. The itineraries explored here consist of key participants' accounts of the city, framed both as a counterhegemonic form of knowledge production and as a narrative device to highlight the interconnections that animate and wound this city. These walks were carried out between October and November 2024, then enriched by further field walks in April and May 2025. The first itinerary was proposed as a *recorrido etnografico*, by walking and driving (four hours) under the guidance of Amine Houari, an architect and member of the ThinkTanger organisation. The second and third itineraries were carried out as walking interviews (three hours each) with Abdou Benattabou, an ecologist, artist, and pedagogist who manages a cultural space in Tangier, and Mustapha[1], a fisherman and inhabitant of the Bab el Bahr neighbourhood. The first itinerary is an established route imagined by ThinkTanger, a cultural agency and urban platform, that is occasionally offered to researchers, artists-in-residence, or interested local people. The second and third itineraries were proposed spontaneously by two key informants to share key representative places of the city's ongoing shifts. These itineraries were then extended, remade and retraced through walking interviews conducted in April–May 2025 with four research participants, namely three fishermen—Anas, Oussama and Mohammed[2]—and Amina Mourid, co-founder of ThinkTanger. Consequently, these itineraries yield an analysis that brings together heterogeneous experiences, including my own situated position as a European cisgender woman collaborating mainly with Moroccan cisgender men. This methodological orientation emerged from the field in a stage of research focused primarily on spatial transformations of food practices in urban natures, rather than on the different layers of foodways themselves, which are addressed elsewhere. Accordingly, participation in these itineraries extended beyond the fishermen who constitute the project's core interlocutors, encompassing Tangier residents engaged in examining urban change from ecological, urban planning, and food-related perspectives.

4 Tangier and its Itineraries

Tanger is a strategic city between Africa and Europe, the Atlantic, and the Mediterranean. It arose from plural civilisations (Amazigh, Phoenician, Roman, Arab) as well as from different phases of European colonisation. Portuguese, British, French, and Spanish layered over Tangier inscribing morphological divides in the city that largely persist to this

[1] Fictional name for anonymization.

[2] Fictional names for anonymization.

day, with a peak presence during the International City period (1923–1956), when Tangier was governed by a council of European diplomats with a minority of Moroccan representatives (Simone, 2023). The historic conflicts between the region and the kingdom, including the Rif War and many consequent revolts, made Tangier, as the whole northern region, an antagonist and an independentist subject, contributing to its exclusion from development plans until King Mohammed VI, the current monarch, ascended to the throne. Due to its economic and political marginalisation, Tangier was entrenched in an imaginary of illegal traffic and smuggling, informality, as well as orientalist charm built on decades of colonial representations. Over the past twenty years, Morocco has undertaken significant structural reforms in various areas, including economic, political, social, and environmental, particularly in the country's major cities. Since 2003, with the launch of the Tanger-Med port construction project, Tangier has experienced large-scale urban development, transforming from a relatively marginalised medium-sized city to the second-largest economic centre in the country and a strategic regional centre (Mourid, 2024). This abrupt urban and infrastructural development clearly had repercussions for the city's social fabric, causing congestion on road infrastructure, inadequate public services, under-equipped neighbourhoods, and uncontrolled construction arising "overnight", often in informal ways (Mourid, 2024; Simone, 2023). Over these past two decades, these processes stimulated internal migration flows within the country and across the continent, which made the city quadruple in size, reaching a population of more than one million. The Tanger Métropole program (2013–2017), launched by King Mohammed VI, aims to rethink the urban plan, enhancing mobility between the Atlantic and Mediterranean coasts, alongside other social, economic, and cultural reforms.

As will be explored throughout the itineraries, these projects extend to the urban outskirts, but all originate from the sea. This is not new in Tangier's history, since its main shifts are historically linked to the sea—from colonial conflicts and conquests to ecological changes and more recent commercial and tourist flows. Looking at the key locations of these transformations, such as the coastal neighbourhoods and the waterfront, it is possible to glimpse a different heritage linked to the sea that goes beyond development and conquest. This is a set of practices and knowledge that are not part of the official narrative of Tangier's heritage but embody a centuries-old relationship between the sea and the city, starting with food. Artisanal fishing in Tangier encompasses ancestral knowledge related to catching, drying, and cooking fish, as well as myths and stories embedded in rocks and beaches around the medina, each of which has a name. Fishers keep this heritage alive through their photographs, fishing tools, and techniques, which are often named using Spanish terms as a trace of decades of competition over the fishing sovereignty in the Strait of Gibraltar, as well as long-standing circulations of ideas and technologies. Despite the city's large port and fish markets—both the old medina's market and the wholesale market—local fish is largely inaccessible for inhabitants due to high prices and export-oriented market dynamics. Traditional fish dishes, such as the sardine tajine or fish broth, can still be found in a few non-touristic restaurants in the old medina, or on the table of artisanal fishers who, despite the increased scarcity of fish along the urban coastline, continue to catch, sell, and share it.

The following itineraries will retrace some of the key places of both the urban development projects and the daily fishing activities. Following the steps and memories of our interlocutors, it will be possible to enter the fractures generated by these transformations, especially in the socioecological dimension and the way it affects food practices. Eventually, the transformations of coastal urban natures and the impacts on the city's artisanal foodways will be analysed.

4.1 First Itinerary: Tanger de dos à la Mer, Rhythms, and Extensions

Tanger de dos à la mer is a non-touristic tour proposed by ThinkTanger, a cultural organisation which investigates the development trajectory of the city from an urbanistic perspective, with an attentiveness towards the spaces of possibilities that can be generated through savoirs citoyens, namely local knowledge (Mourid, 2024). The proposed itinerary begins from the sea, albeit turning its back on it, challenging the prevailing narrative of the city as oriented towards the northern shore of the Mediterranean and represented by outsiders' gaze. Therefore, the tour aims to unravel urban extension, establishing connections between three central themes for contemporary Tangier that make visible the intersection of spatial and food injustice. The first, namely Tangier's relationship with the sea from an infrastructural perspective, can be traced starting with the port (Fig. 1).

The artisanal fishing port is key to the transformation of the urban waterfront, as part of the Tangier Métropole development plan, closely

Fig. 1 First itinerary map: Tangier Back to the sea. ThinkTanger

connected with the construction of the Tanger-Med port (Mareï & Wippel, 2020; Wippel, 2022). Until 2013, the urban port incorporated commercial, tourist, and fishing functions. The site of the former artisanal fishing harbour is now home to the luxury Marina Tanja Bay project, co-designed by a national agency and a company based in the Emirates. The routes of tourist ships, except for the ferry from Tarifa, have been transferred to Tanger-Med, along with the landing of commercial container ships, whose traffic is progressively growing. The port is guarded by three different security forces, including the military. Several fences separate the port from the outside, including a railing over which fish restaurants are located.

> These restaurants are not inviting; I would not be comfortable eating here. Do you know why? It is the smell. It is not the smell of fresh, local fish. It is a mixture of fish, rubbish, and waste oil. It is not like in Essaouria, where you can buy fresh fish directly from the fishermen and eat it at the harbour. A few years ago, it was like that here too. (Amina Mourid, April 2025)

Bordering the port, focusing on embodied sensations such as smell or the feeling of being out of place given by the securitisation of the perimeter, stimulates the memory of the old port as an open and convivial place, where the fishing community was in dialogue with the city. Similarly, the sense of disgust generated by low-quality fish restaurants raises questions about the accessibility of local fresh fish, prized for their high quality and exceptional taste, which is progressively scarce and primarily directed at the foreign market.

The second dynamic emerging from the urban tour related to the relationship with the sea is the progressive concreting over the shoreline. An exemplary location of this process is Ghandouri (Fig. 2).

The Ghandouri area is a traditional urban fishing spot. Its mirador is an emblematic site in the area, a vantage point where people enjoy the view and access the seaside for walking and fishing. Around the mirador, on the edge of the sea, several construction sites are being developed to create an affluent residential and tourist area. Walking past the construction sites reveals construction waste seeping into the sea and signs of erosion along the unstable rocky shoreline barrier. This construction site is emblematic of a broader process of coastal privatisation, through which the shoreline is rendered progressively less accessible and ecologically viable for small-scale subsistence fishing. Artisanal fishers are experiencing a progressive

Fig. 2 A fisherman standing on a rock facing the port, with the construction site at his back. View from the Ghandouri mirador, October 2024. Author's photo

displacement, starting from the historical fishing areas connecting the medina and the old port and now extending to more peripheral areas, such as Ghandouri. Nonetheless, smaller groups of fishermen continue to fish with their rods standing on nearby rocks or to hunt octopus, primarily for subsistence. These informal fishers usually sell their catch to people passing by, or at the fish market in the old medina.

Walking along this stretch of coastline with people holding different perspectives revealed multiple critical dimensions of the area. On one hand, the perspective of practitioner and architect Amine Houari highlighted the material and structural impacts of construction in this zone. Indeed, a significant amount of concrete is needed to stabilise the sandy ground, which consistently erodes the coast. On the other hand, walking with the fisherman Oussama made tangible elements overlooked in urban planning that are, in fact, crucial for artisanal fishing. These were primarily sensorial factors, such as intense lighting, car noise, and construction materials that pollute the sea and force fish populations to migrate. The progressive construction development, which in the last ten years involved the eastern coast of the city, has gone hand in hand with the erosion of the Mediterranean scrubland. The Charf Hill top, another key place encountered in the first itinerary, offers an overview of the urbanisation of the city over the last twenty years, between the north-eastern coastal area populated by large commercial and residential buildings, in stark contrast to the area developed in the south-west, namely a mosaic of informal constructions, regularised after a few years.

The third process unfolded through the tour concerns regional and global networks of value flows, of which the industrial area and Tangier Free Zone are emblematic. Inaugurated in 1999, the TFZ is located about 15 kilometres from Tangier in a tight connection with the Tanger-Med port, the Tangier international airport, and the highway networks, in order to accelerate commodities flows (Mourid, 2024; Wippel, 2022). With 500 hectares dedicated to industrial, logistics, and commercial activities, the TFZ hosts 900 companies, including the automotive industry (e.g. Renault), aerospace, textiles, electronics, and services. To cross this area, it was necessary to drive along the Roquade des deux Mers, a further infrastructural work of the last decade, key to the mobility of food commodities between the Atlantic and the Mediterranean. The wholesale fish market is located here and contains products coming from the two shores of Tangier, which are then sold to intermediaries. Driving through this area provides an overview of the social spin-offs of the TFZ project, i.e. a large-scale social housing scheme to accommodate new residents attracted by the area's job opportunities (Wippel, 2022). The functionalist structure of these neighbourhoods, namely pure concrete disconnected from the social and cultural fabric of the city, worries Amine: "We are repeating the same mistakes as Europe" (Amine Houari, October 2024). The outcome of this urban sprawl is a high risk of urban segregation for the new inhabitants, coming from the rural northern region as well as from across the country. Furthermore, the TFZ is a major producer of industrial waste. Fishermen, observing these shifts from the coast, associate the progressive urbanisation and the massive development of the industrial zone, along with the mismanagement of industrial waste, with the pollution of the sea and urban waterways, impacting the urban fish population.

4.2 *Second Itinerary: More-Than-Human Entanglements*

The second itinerary will unfold the co-production of urban natures from a more-than-human perspective. The following steps of this walk will trace the co-dependence of fishing practices with key ecotones, such as waterfront vegetation, wetlands, and waterways. Focusing on overlooked more-than-human relations—such as the grass-sand-fish entanglement—it will be possible to highlight the erosion of urban natures as a result of recent urban development projects. At the same time, the itinerary will show how both human and non-human inhabitants react to these shifts, reinventing their use of space and their food foraging strategies (Fig. 3).

The removal of native flora is to be traced in overbuilding and urban sprawl, but also in a purposeful vision of landscape architecture, which foresees the waterfront as crucial for the neoliberal and global city narrative, connecting Tanja to the Mediterranean as well as to the Gulf (Simone, 2023; Wippel, 2022). Watercourses, from the Oued Lihoud (Jewish River) to the Oued el Maleh (salty river), to the western water frontier of Cap Spartel, are often interrupted and reshaped according to urban planning needs. Ecologically, this planning model hinders the circulation of diverse more-than-human agents, disrupting the biological entanglements through which they communicate underwater and underground—circulations that are crucial for sustaining the balance between marine and terrestrial ecologies (Temsamani, 2025). The introduction of exotic species along the waterfront as part of the beautification driven by the Tanger Métropole plan further compromises this mutuality, as in the case of the Kikuyu Grass, an invasive species from Kenya (Houari & Hršak, 2023). The Kikuyu can climb over other plants, shading them out and producing herbicidal toxins that kill competing species. It prevents the growth of new shoots of other plants, can kill small trees and can choke ponds and waterways. Like other tropical varieties transplanted to the corniche, it is watered daily and pruned very frequently, exemplifying the canvas function of the waterfront on which the metropolis and the monarchy can project their vision of urban regeneration and modernisation. Indeed, in the past twenty years, the same lawn has been used for regenerated and institutional spaces of Rabat and Casablanca. While walking along the waterfront with Mohammed, a recreational fisherman, he says:

"There is no life in this grass; it leaves no space for anything. There is not an insect, not a single spontaneous plant along the waterfront, except

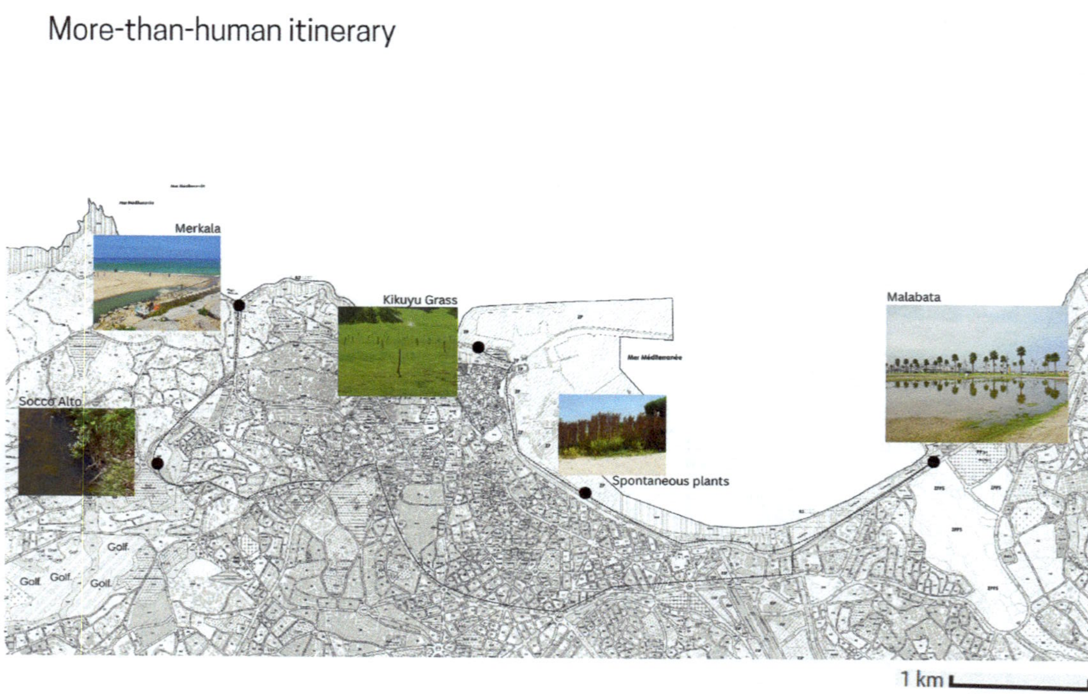

Fig. 3 Second itinerary map. Author's materials

for some tiny marginal spaces. Because of administrative and ownership gaps, given the different projects going on, the local vegetation has remained at these thresholds. You can recognise it by the smell. Here, for example, there are wonderful yellow flowers in spring, and you have to be careful not to step on snails. They have come to this small space on the sidelines, as they cannot survive in the grass" (Mohammed, May 2025).

The biodiversity hosted by local vegetation on the waterfront, in Mohammed's view, is crucial to nourish sand and, consequently, fish. The sand on the municipal beach, which hosted various mussels and sea worm species that nourished both humans and fish till a few decades ago, is now "dead", in his words.

Another place crossed in the more-than-human walk along the waterfront and crucial for socioecological relations is the Malabata wetland. Walking through Malabata, it appears as a hilly green space next to the seaside. The hills, however, are the product of years of concrete waste used in various phases of the Tanger métropole project, including the construction of the waterfront. Malabata is a key ecological area of the city, namely a wetland generated by the encounter of the sea and the watercourse streaming from the Rif mountains, the Oued el Maleh. Walking through this space, several traces of ecological erosion and shifting functions can be followed: old wooden fences, indicating former semi-abandoned farmlands, bricks, coastal plants, as well as camel and birds' footprints. The avian population in Malabata has undergone a substantial decline, which is of particular concern given its role as a refuge for birds during their transcontinental migration. Walking among the bricks—evidence of the concrete cemetery underground—also salicornia, a typical coastal plant, can be found. While traditionally harvested for eating and medical purposes, salicornia is progressively disappearing from the area. Despite the hindered access to this space and its ecological decline, inhabitants are still making diverse uses of it, namely collecting worms for fish bait, grazing small groups of cattle, and organising convivial moments such as picnics.

Indeed, the management of space is in a bureaucratic vacuum between the ministries and the metropolitan city, making it home to various informal and peripheral practices (Simone, 2023) (Fig. 4).

Another crucial wetland that has been progressively lost was generated by the Oued Lihoud, a river located in a former fertile agricultural area, currently known as the California neighbourhood. Elevated above sea level, the neighbourhood was inhabited by French, Spanish, and

Fig. 4 Construction sites next to Malabata. May 2025. Author's photo

English settlers, who domesticated the wetland by importing and transplanting various eucalyptus trees during the Tangier international era. Mohammed, reflecting on his childhood memories, remembered the river as an ecological meeting space:

"I used to go on picnics here with the family, and bathe in the river. At school, we would go out with the teachers here to learn biology, looking at the leaves, the animals...now nature is no more". (Mohammed, May 2025).

The neighbourhood, as the upmarket name suggests, is still inhabited by middle and upper classes, both national and international. Among the neighbourhood's various amenities is a large golf and horse-riding area. Of the river, channelled underground a few kilometres further north, a thin stream remains, where a dense colony of five hundred turtles lives among various urban wastes. Meeting the turtles and visiting the former river were central aims of Abdou's itinerary. Through this encounter, he sought to share the marginalisation of non-human life in contemporary Tangier, as well as the quiet forms of resistance these agents embody in a city that has pushed them to its edges. The stream's water flows into the drainage channel from the neighbourhoods behind and the industrial area, which is then channelled to Merkala, one of Tangier's key beaches. Merkala has historically been a place for artisanal fishing, but it also embodies symbolic and ritual values for the inhabitants of coastal western neighbourhoods. A place of purification and connection with the sea in traditional knowledge, Merkala is now heavily contaminated, as pointed out by some of the interviewees complaining about the smell while walking by.

4.3 *Third Itinerary: Bab el Bahr, Spatial and Social Fragmentation in Fishers' Communities*

This itinerary brings us back to the sea, enriched by the information and sensations gathered during the previous walks. It follows the traces of spatial and socioecological change in the coastal neighbourhoods facing the waterfront, unfolding the connection of inhabitants with the sea as manifested in artisanal fishing. These neighbourhoods are called Bab el Bahr (Door of the sea), Mershan (the neighbourhood of merchants), and El Hafa (the margin). Walking along the edge of Bab el Bahr, at first sight, its hard to make out the remnants of demolished houses. In 2018, some informal self-built dwellings were identified as unsafe by the Ministry of National Land Use Planning and thus demolished. Some of the site's residents were relocated to peripheral industrial neighbourhoods such as Mesnana, next to the TFZ industrial zone, which is two walking hours away from the coast. Mustapha, while looking at the house rubble and smelling the salty breeze, says: "I am like a fish. If you take me far from the sea, out of here, you will kill me" (Mustapha, November 2024). With the forced relocation of part of this community, the solidarity networks and the "atmosphere" within the neighbourhood were affected, in many fishermen's words, making room for significant speculation by investors in the tourism sector. Along with houses, some collective food infrastructures have also been demolished, such as the collective oven, now replaced by a parking lot. The oven was a key meeting point for the neighbourhood and a shared resource, where families brought their fish tajine or bread to bake. A relevant portion of local vegetation in this area was also eradicated, including bamboos, traditionally used as rods for fishing on Sidi Boughadif, the beach situated just beneath the bastions now covered by concrete. Inhabitants had long fished there for livelihood, for sharing within the neighbourhood, as well as for the conviviality, gathering the community around tajines cooked collectively on the shore. Convivial moments of this kind are currently inaccessible: the concrete has replaced the sand, and direct access to the sea is interrupted by the high-speed boulevard. Although the corniche is promoted as the infrastructure reconnecting the city to the sea, walking along its inner side quickly reveals that it is designed far more for fast-moving cars than for pedestrians, with safe crossing points almost impossible to find. Fishermen often remarked on the constant noise of the cars while walking along this road, seeing it as harmful to the fish and unsettling for themselves. They explained that much like the fish, they seek stillness and silence, approaching fishing as an intimate, introspective ritual. Small-scale fishing is still practised on the rocks bordering the waterfront, but fish are gradually disappearing because of the pollution brought by the intense building in the area. While walking down the neighbourhood and reaching the seaside, we passed by the remains of some houses, wrapped in the green Kikuyu grass facing the highly surveilled waterfront (Figs. 5 and 6).

5 Relational Fishing Practices in Contested Urban Places: A Trans-Scalar Analysis

These itineraries questioned urban foodways from an unusual perspective. Starting from the spatial transformations of key places related to urban marine ecology, it was possible to frame the trans-scalar dynamics impacting fishing practices. These encompass, as shown in the first itinerary, rapid industrial and commercial development of the region, which makes Tangier an expanding hub of commodity flows, including local fish, which is progressively redirected to the foreign market. From these shifts, a progressive concretisation of the city emerges, but also an impressive rebranding of urban spaces such as the waterfront (Wippel, 2023). As a result, general privatisation of the coast and eviction of informal and artisanal food practices from key urban spaces can be noted, either directly or indirectly. The accessibility to the sea for the inhabitants of coastal districts has emerged as both a matter of spatial and food justice, questioning how the displacement of fishers and fish from the coast undermines the food sovereignty of coastal communities and their access to culturally compatible food practices. At the same time, several processes polluting the sea, also connected to industrial expansion, are hindering the survival of the littoral ecotone and indirectly preventing artisanal

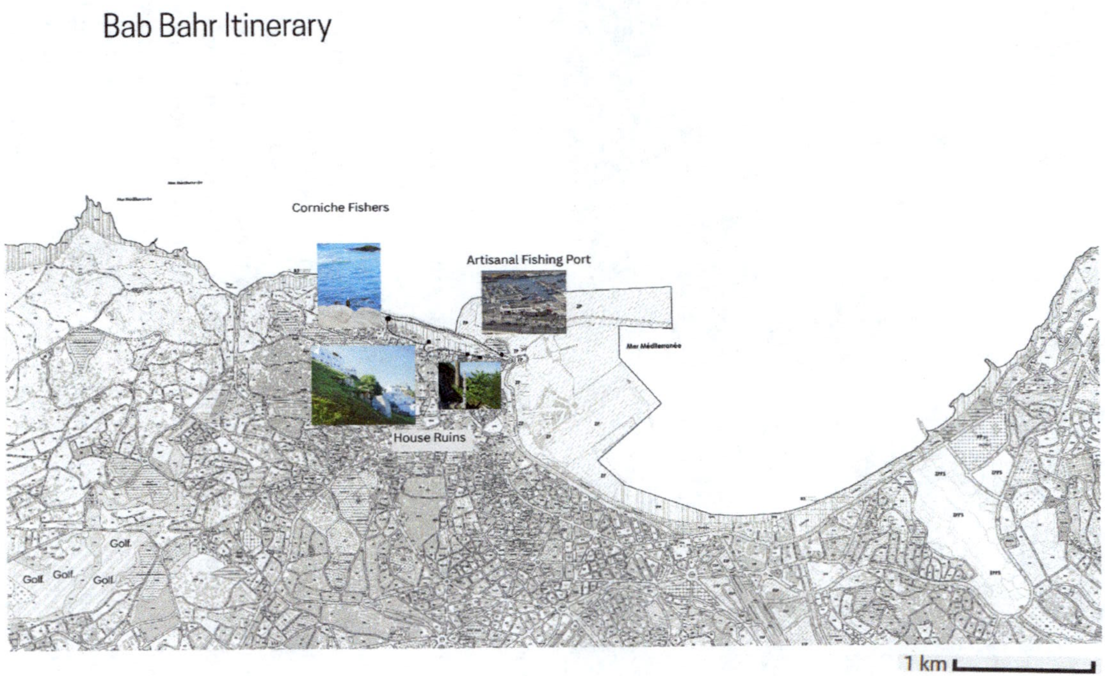

Fig. 5 Third itinerary map. Author's material

Fig. 6 Remains of demolished houses. October 2024. Author's photo

fishers from their practices. This loss can be sensed by the absence of key non-humans along the coast, such as mussels, seaweeds, and worms—all actors entangled with marine foodways in material and symbolic ways. Experiences and memories shared by fishermen while walking in places such as Ghandouri and Bab el Bahr resonate with Poe's findings, namely that small-scale foraging and fishing practices can be seen as relational ecologies that embody a sense of belonging to these places (2014). While walking together, participants manifested this relation by sharing their knowledge of changing places and ecologies. They recalled memories about fishing practices sparked by a smell or a transformed landscape, expressing both feelings of affection and nostalgia, but also a sense of injustice and dispossession. In these coastal neighbourhoods, fishing has been the pivot of the community and the first connection to the sea, a way of being in place and an active part of the city's meaning-making. Observing these connections from the perspective of spatial transformation allows us to highlight the progressive loss of key locations embodying this relationship mediated by food, such as the community oven, the neighbourhood beach, and the port. The artisanal fishing port, once perceived as an open space of encounter, is now depicted as a blind spot on the map. Closing its gates, in the fishermen's view, meant hindering the relationship between citizens of Tangier and a relevant part of their food heritage expressed by direct selling and bargaining, as well as eating fresh catches directly at the port. At the same time, the militarisation of the artisanal fishing port contributes to the stigma of fishing as a potentially clandestine economy related to migration and smuggling. The embodied feeling of rejection while walking along the fences of the port (and within the port, in the case of fishermen) is to be discussed while taking into account contextual political nuances. To understand Tangier's waterfront—both its spatial transformation and changing fishing practices—it is necessary to face the northern shore of the Mediterranean and consider how the moral authority of the border is impacting the everyday life of fishermen. Here, crucial new questions emerge about the positionality of Tangier as a Southern Mediterranean City, leading us to interpret it not as a place passively built in reflection of the northern shore, but as a city that mobilises the Mediterranean to its advantage in the construction of a peculiar national discourse of modernity. Nonetheless, alternative and frictional narratives about the city can emerge from overlooked practices, such as fishing among construction debris in Ghandouri or next to the port fences, as well as foraging salicornia or collecting worms in Malabata. These practices exceed the institutional narrative of Tangier's relationship with the sea and open spaces for other ways of inhabiting the city. Despite the moral registers of authority projected in the waterfront space, the coexistence with these practices allows the emergence of what Simone defines the Southerness of Tangier, namely "the possibility of things being together without the hint of a system or framework, that things can be with each other without being subsumed into an imaginary of a larger totality [...] in a landscape that is both useless and makes use of everything" (Simone, 2023, p. 357).

Walking while taking photos allowed the researcher to attune with different rhythms of the city and to get in contact with the more-than-human subjects, such as small turtles hidden in the rubbish, octopus on the rocks, and seaweeds on the shore. This attentive observation was mediated by the interlocutors during the walks. Their guide, orienting the filming process, allowed an "inhabitant-led gaze" to be archived through the collection of visual materials. Taking pictures while walking with participants—focusing on subjects, places, and details of their interest—can be considered as a caring and engaged observation, contesting the more conventional flâneur attitude of the urban walks. All the participants are engaged in documenting the loss of urban nature and of ecological relationships and food sovereignty in these spaces, in different ways. It is noteworthy that each of the participants proposed to walk as a way of sharing, namely to follow the traces, the temporalities of these changes, and to archive them together. The photographs presented here are to be understood as tiles of a mosaic made by diverse and entangled gazes gathered throughout these itineraries. These visual materials must not be interpreted as a neutral representation of what Tangier is, but instead as a positioned perspective developed in this specific space of knowledge production. Furthermore, the stories shared through these itineraries are more than visual: many segments of these walks highlighted sensuous information about places, such as their sounds, smells, textures, and atmospheres. All these traces have been key both to attune to place and to trace the memory and sense of belonging that guided the walks. Starting from embodied experiences of ecological and spatial change, this approach facilitated a trans-scalar analysis of the neoliberal development of the waterfront and the tightening of its border, as well as the impact of this planning model on urban natures, more-than-human agents, and fishers' communities—all entangled in the explored fishing places.

6 Conclusions

These itineraries have traced plural connections between different actors and processes involved in spatial and food injustice, in a city which is undergoing profound transformation from the sea to its outskirts. These processes have been observed at different scales, namely at the regional, metropolitan, and street levels, and then analysed according to historical and contemporary political and ecological pressures. Artisanal fishing in these liminal places proves to be a relational ecology of belonging, as it uncovers lively knowledge about the transformation of socionatures connecting inhabitants to the ocean. The spatial change of neighbourhoods and coastal areas—as well as the ecological erosion of waterways, wetlands, and the ocean—cannot be understood as separate from the everyday life of Tangier's inhabitants and their food cultures. On the contrary, key fishing places—where fish is caught, sold, shared, and cooked—appeared as a knot of socioecological relations and local knowledge which is key to sustaining local marine foodways. More significant than a remnant, artisanal fishing is a vibrant archive of urban shifts, recording the loss of coastal space due to privatisation and pollution, but is also a practice of urban food sovereignty in a context where local fish is mainly exported or inaccessible. Small-scale fishermen possess an ecological knowledge of the city, witness changes in emplaced foodways, and interfere in urban waterfront planning.

Walking through fish places together with the participants led to rich discussions and memories elicited by sensorial elements—such as the smell of indigenous vegetation or fish restaurants, the discomfort of bordering the surveilled fences, and the car noise along the corniche. From a methodological standpoint, these itineraries aim to unfold the city to its extension while grounding the understanding of key urban natures and food places through an embodied and attentive spatial exploration. The combination of participative and "slow" geographical observation with a trans-scalar analysis of political and ecological processes is presented here as a key strategy to unfold knowledges, relations and practices embedded in urban food places. The multimodality of the tools involved, such as photography and itinerary mapping, aims to make the exploration more immersive and accessible. While mapping the sites of transformation often appears as a static representation of a fixed scale, itineraries offer a mobile narrative of the city, dotted with images that bear witness to the fragment of time between what still exists and what is about to change in Tangier's shifting space and identity.

Advancing reflections on the sensuous exploration of food, this paper proposes to embody foodways not only by eating, but also by experiencing the different places and agents that shape food as an urban subject. Places where artisanal fishing is practised can open up space for more relational understanding of food. Overcoming the analysis of food as a commodity that is sold and consumed, food can be a lens to explore meaning-making practices rooted in local ecologies and urban knowledges that diverge from the neoliberal models and values of inhabiting the city. By smelling, touching, photographing, and mapping these emplaced entanglements, it is possible to reflect on the connections and fractures that characterise Tangier's marine foodways, opening space for further research on its material and spiritual dimensions.

References

Alkon, A. H. (2013). The socio-nature of local organic food. *Antipode, 45*(3), 663–680.

Barchetta, L. (2019). *Riverbanks made by walking: understanding the temporalities of urban nature through atmospheres* (PhD Thesis). Gran Sasso Science Institute: Urban Studies Doctoral Programme.

Barron, A. (2021). Photo go-alongs. In A. Barron, A. L. Browne, U. Ehgartner, S. M. Hall, L. Pottinger, & J. Ritson (Eds.), *Methods for change: Impactful social science methodologies for 21st century problems*. Aspect and The University of Manchester.

Barua, M. (2014). Bio-geo-Graphy: Landscape, dwelling, and the political ecology of human-elephant relations. *Environment and Planning D: Society and Space, 32*(5), 915–934. https://doi.org/10.1068/d4213

Barua, M., & Sinha, A. (2017). Animating the urban: An ethological and geographical conversation. *Social & Cultural Geography, 20*(8), 1160–1180. https://doi.org/10.1080/14649365.2017.1409908

Brooks, A., & Hubbard, P. (2022). Oysteropolis: Animals in coastal gentrification. *Environment and Planning E: Nature and Space, 5*(3), 1495–1517. https://doi.org/10.1177/25148486211024901

Bunce, S., & Desfor, G. (2007). Introduction to "political ecologies of urban waterfront transformations". *Cities, 24*(4), 251–258. https://doi.org/10.1016/j.cities.2007.02.001

Counihan, C. (1999). *The anthropology of food and body: Gender, meaning, and power*. Routledge.

Cresswell, T. (2006). *On the move. Mobility in the modern western world*. Routledge.

Duruz, J., Giovanangeli, A., Loda, A., & Manganas, N. (2022). Emotional scapes in Mediterranean port cities: Walking Barcelona, Marseille and Genova. *European Journal of Cultural Studies, 26*(4), 572–597. https://doi.org/10.1177/13675494221141921. (Original work published 2023).

Finlay, J. M., & Bowman, J. A. (2017). Geographies on the move: A practical and theoretical approach to the Mobile interview. *The Professional Geographer, 69*(2), 263–274. https://doi.org/10.1080/00330124.2016.1229623

Gandy, M. (2024). Attentive observation: Walking, listening, staying put. *Annals of the American Association of Geographers, 114*(7), 1386–1404. https://doi.org/10.1080/24694452.2024.2353841

Goodman, M. K. (2016). Food geographies I: Relational foodscapes and the busy-ness of being more-than-food. *Progress in Human Geography, 40*(2), 257–266.

Grove, K. (2009). Rethinking the nature of urban environmental politics: Security, subjectivity, and the nonhuman. *Geoforum, 40*(2), 207–216.

Hammelman, E. C., Levkoe, C. Z., & Reynolds, K. (2024). *Radical food geographies: Power, knowledge and resistance*. Routledge.

Heynen, N. (2018). Urban political ecology III: The feminist and queer century. *Progress in Human Geography, 42*(3), 446–452. https://doi.org/10.1177/0309132517693336

Holton, M., & Riley, M. (2014). Talking on the move: Place-based interviewing with undergraduate students. *Area, 46*(1), 59–65.

Houari, A., & Hršak, L. (2023). *Fill in the gap: Tanger Glossary*. https://bureaulada.com/tanger-glossary

Lee, J., & Ingold, T. (2006). Fieldwork on foot: Perceiving, routing, socializing. In S. Coleman & P. Collins (Eds.), *Locating the field: Space, place and context in anthropology*. Routledge.

Lorimer, J. (2010). Moving image methodologies for more-than-human geographies. *Cultural Geographies, 17*(2), 237–258. https://doi.org/10.1177/1474474010363853

Mareï, N., & Wippel, S. (2020). *Une perspective urbaine de la régionalisation du monde: Tanger, métropole (eur) africaine* (p. 4). Revue belge de géographie.

Mason, O., Sarma, J., Sidaway, J. D., Bonnett, A., Hubbard, P., Jamil, G., Middleton, J., O'Neill, M., Riding, J., & Rose, M. (2023). Interventions in walking methods in political geography. *Political Geography, 106*, 102937. https://doi.org/10.1016/j.polgeo.2023.102937

Merriman, P. (2014). Rethinking Mobile methods. *Mobilities, 2*(9), 167–187.

Montefusco, G. (2026). Relational food geographies, underwater: Discussing food and Oceans' entanglements along urban waterfronts. *Geography Compass, 20*(4). https://doi.org/10.1111/gec3.70074

Moragues-Faus, A., & Marsden, T. (2017). The political ecology of food: Carving 'spaces of possibility' in a new research agenda. *Journal of Rural Studies, 55*, 275–288. https://doi.org/10.1016/j.jrurstud.2017.08.016

Mourid, A. (2024). *De l'initiative citoyenne à la gouvernance urbaine: Think Tanger au cœur des mutations de la ville* (Master's thesis). IUAR Aix-Marseille.

Pierucci, G. (2025). *Urban Wildness Frontier. Ethnogeographies and Mobility of the Unplanned Green in Rome and Granada* (Ph.D. Thesis). University of Padua.

Poe, M. R., LeCompte, J., McLain, R., & Hurley, P. (2014). Urban foraging and the relational ecologies of belonging. *Social & Cultural Geography, 15*(8), 901–919. https://doi.org/10.1080/14649365.2014.908232

Rodenbiker, J. (2023). Urban oceans: Social differentiation in the city and the sea. *Environment and Planning E: Nature and Space, 6*(1), 412–432. https://doi.org/10.1177/25148486221078690

Sheller, M. (2018). *Mobility justice: The politics of movement in an age of extremes*. Verso.

Sheller, M., & Urry, J. (2006). The new mobilities paradigm. *Environment and Planning A, 38*(2), 207–226.

Simone, A. (2023). Making use of everything: Tangier and its Southern, peripheral practices. *South Atlantic Quarterly, 122*(2), 339–359.

Streule, M. (2020). Doing mobile ethnography: Grounded, situated and comparative. *Urban Studies, 57*(2), 421–438. https://doi.org/10.1177/0042098018817418

Swyngedouw, E. (1996). The city as a hybrid: On nature, society and cyborg urbanization. *Capitalism Nature Socialism, 7*(2), 65–80.

Temsamani, M. (2025). *Tanger, naturellement! Connaitre et protéger la nature en ville*. Tangier. ISBN: 978-9920-23-147-3.

Tzaninis, Y., Mandler, T., Kaika, M., & Keil, R. (2020). Moving urban political ecology beyond the 'urbanization of nature'. *Progress in Human Geography, 45*(2), 229–252. https://doi.org/10.1177/0309132520903350

Vacchiano, F. (2022). *Antropologia della dignità: Aspirazioni, moralità e ricerca del benessere nel Marocco contemporaneo*. Ombre corte.

Wilson, H. F. (2024). Urban maintenance as compromise: Coming to terms with the multispecies city. *Environment and Planning E: Nature and Space, 8*(1), 232–250. https://doi.org/10.1177/25148486241280354. (Original work published 2025).

Wippel, S. (2022) Entre commerce et transport, loisirs et plaisirs: les reconfigurations et les mondialisations du front de mer tangérois. In Mareï, N., Florin, B. et Iraki, A (eds) *Le Nord du Maroc. Intégration, inégalités et résistances*. Les Cahiers d'EMAM (34), Tangier.

Wippel, S. (2023). The multilevel branding of Tangier at spatial and temporal interfaces. In S. Wippel (Ed.), *Branding the Middle East: Communication strategies and image building from Qom to Casablanca* (pp. 429–458). De Gruyter. https://doi.org/10.1515/9783110741100-026

Separation as Oppression? Rethinking Gendered Power Dynamics Through Ocean Foodways in Southern Morocco

Beatrice Ferlaino

Abstract This chapter examines how gender roles and spatial separation are enacted through everyday food practices in Sidi Bounouar, a rural Amazigh village on Morocco's southern Atlantic coast. Based on ethnographic fieldwork, the study explores relations of belonging and power between women's and men's groups within food production, transformation, and consumption, in a context of ecological degradation that has led to the abandonment of subsistence agriculture and an increasing reliance on sea-based livelihoods. Tourism and external development interventions are reshaping the local economy while introducing "gender" as an exogenous category, deployed to render the village legible to Western and institutional gazes that shape dominant gender mainstreaming approaches within development and aid programmes. However, the research demonstrates how everyday food-related activities illuminate the complexity of gender relations, challenging taken-for-granted power dynamics and reconfiguring the meanings of spatial separation. Rather than reproducing

B. Ferlaino (✉)
Department of Cultures, Politics and Society, University of Turin, Turin, Italy
e-mail: beatrice.ferlaino@unito.it

P. Minoia (ed.), *Foodways As Sites of Knowledge and Transformation*, Emplacing Food, https://doi.org/10.1007/978-3-032-28126-5_5

depictions of rigid and segregated gender roles around food, the chapter aligns with non-white and Islamic feminist scholarship in arguing that gender separation does not inherently signify inequality, but may instead function as a form of community social cohesion sustained through situated gender relations. In a degraded coastal environment, practices of fishing and food preparation emerge as sites of gendered intersection, as fish, shellfish, and meals circulate across groups, forging connections between the shoreline and domestic kitchens.

Keywords Gender · Power dynamics · Foodways · Fishing/foot-fishing spaces · Morocco

1 Introduction

> When the tide rises and women return to the beach, they change their clothes to dry, gather shrubs to make tea, and they arrange the mussels in a circle, side by side, closed upwards and open downwards, directly on the beach. They place the wood on top to be burnt and light a fire for partial cooking and smoking. The wood that are used to set fire are a mix of dried local shrubs: ifarskil, akjboud, and a succulent plant called darmouss. In some villages, they also add argan wood to the fire. No woman may start the fire in her own circle of mussels if the others have not finished theirs. If one finishes first, she will give the others a hand. The fire lasts for 5-10 minutes. After the fire has been extinguished, it is left for a few minutes to cool down. At this point all the ash residues are eliminated with a brush and the wind made with a cardboard. Then, the mussels are taken one by one, opened, and extracted with the help of the shell, as if it were a spoon, and laid out on a clean fabric in the sun. Mussels dry in the sun 1-2 days (depending on the intensity of the sun). When mussels become hard and break easily, they have reached the correct dryness. At this point, they are put in fabric bags in a cool, clean and isolated place. (Fondazione Slow Food, 2025)

This is how the international association Slow Food narrates the process through which *Wakkad* mussels are produced by the women of Sidi Bounouar, a 200-person Amazigh village in the Souss Massa region (100 km south of Agadir) along the Atlantic coast of Southern

Morocco. Presented as a local tradition and increasingly less practiced in the everyday life of the community, this technique historically emerged from the need to transport mussels from the beach to inland villages, lightened by removing their shells. Today it is framed as part of the region's cultural heritage: documented in a short film—entitled *Taghart* (Terras, 2024)— and formalised as a Slow Food Presidium product, a project that, according to the association, "protects traditional and artisanal products that are at risk of disappearing" (Slow Food, 2025) by codifying them through written protocols. The name of this Presidium is Wakkad Tigri Mussels, where *Tigri* in the local Amazigh language refers both to shellfish and to the seasonal period dedicated to their harvesting. Gender is presented as a central dimension of this practice: it is through mussel harvesting and processing that the women of Sidi Bounouar are publicly represented, including in several development programmes, which describe them as mussel harvesters. Their access to these initiatives is mediated through the Sahel Aglou Women's Cooperative for Seafood Products, a cooperative with the mandate of valorising women's knowledge related to mussel harvesting and engaging with blue-economy initiatives, which present frequently an explicit focus on the gender dimension.

In this context, gender operates as an exogenous category to be used in international aid programmes to define priorities in project frameworks. The local population, in turn, mobilises this label to align with institutional expectations and to secure access to funding programmes. The women-centred narrative that accompanies this practice captures only a partial dimension of the region's fishing realities, omitting the multiple nuances of shellfish harvesting and the shared knowledge that underpins this work. To fit this activity within development frameworks, the social organisation of Sidi Bounouar is essentialised: simplified into a gender-based separation that sharply divides men and women and portrays women as immutable stewards of local traditions, supposedly uncontaminated by capitalist interferences, a process also documented in other contexts (e.g. (Rodriguez, 2001) (Adolfsson & Madsen, 2020), (Samarasinghe, 2014), (Adusei-Asante, Hancock & Oliveira, 2015). Challenging this essentialism is crucial to abandoning colonial legacies in the representation of non-Western societies. This requires attending to the nuances and interconnections among genders, examining how power, knowledge, and belonging circulate between these groups, and moving beyond opposed binaries such as modern/traditional and gendered power/

oppression. In response to this gender essentialism, and while informed by non-white feminist literatures, this research does not adopt pre-established theoretical frameworks on gender, but rather an ethnographic approach in which analytical categories emerge from everyday practices and relations in the village; men and women are thus discussed relationally rather than as fixed or oppositional groups. Within this field, the gender dimension—and more specifically the reflections developed in this chapter—did not constitute an a priori focus but took gradually shape during my early visits to Sidi Bounouar (more than ten years ago) and further consolidated through repeated returns within the Food and Community Empowerment (FoCE) research project.

The context of Sidi Bounouar offers this opportunity, unsettling assumptions about what the women of the cooperative actually want. During a focus group conducted within the FoCE project with members of the women's cooperative, they articulated an alternative narrative regarding their roles and aspirations. As I noted in my field diary during that meeting,

> They would like to move beyond their role as mere shellfish harvesters. They hope to enter the market as tailors or cooks. [...] Seashells are becoming rarer, marine life is decreasing, and more and more people rely on sea products in a context of unemployment and economic hardship. Harvesting is demanding and dangerous. Many women associate this activity with poverty, and although it remains a fundamental part of their sense of belonging, they want harvesting to be a choice, not a duty. They seek access to other forms of work and markets. They want to diversify their income and create new ways for entering local markets.

Although mussel harvesting has long been tied to women's roles in securing food and is still practiced by many of them, some women now associate it with poverty and necessity, recalling periods when they were solely responsible for feeding the family while men migrated for work. This memory, combined with the increasing scarcity of marine life, pushes women to seek to diversify their income through activities unrelated to fishing. Moreover, the emphasis on "women" aligns more with external expectations shaped by international frameworks than with local reality; shellfish gathering is, in fact, an activity carried out by both men and women, which makes it difficult to identify a sharp gender distinction in either activities or knowledge.

The women's cooperative works together with another important actor in local civil society, the Amoud Association, composed primarily of men from the village. The women's cooperative and the men's association collaborate in the co-design of projects. The desired changes expressed by the women are progressively taking shape: the women's cooperative is now involved not only in fishing-related initiatives but also in tourism, food processing, and other income-generating activities, reflecting a broader trend of using development projects as a pathway into the capitalist economy (Klenk, 2004). This challenges the idea of women as passive, static subjects who merely respond to external pressures, and instead opens space for more nuanced analyses of what it means to be a woman in the context of Sidi Bounouar. More broadly, this invites reflection on how to approach the gender dimension in non-white and Muslim-majority contexts, where women and men are often framed as groups linked by structurally imbalanced power relations—women portrayed as victims, men as enforcers of power, and development agencies as the actors that "redeem" such inequalities.

Spatial separation and spatialised relations have long been used in feminist geography to access gendered power dynamics and women-men relationships in non-Western and Muslim-majority. Since "spatial relations are also power relations" (Murdoch, 2006, p. 23), focusing on space and spatially situated interactions can sustain counter-narratives that challenge dominant representations of gender interactions. This perspective enables a more complex and tangible understanding of gender relations by focusing on their everyday manifestations, thereby avoiding essentialisation. Several studies have drawn on spatial analysis to examine intersections of gender, age, and class by analysing, for instance, how Nigerian women experience public and private spaces differently (Robson, 2000); by exploring dynamic power relationships within Pakistani families (Ajaz, 2022); and by challenging binary categories such as modern/traditional, secular/religious, public/private, and male/female through the everyday experiences of women in Tehran (Bagheri, 2014). In Morocco, similar approaches have been adopted to study urban spaces (Newcomb, 2006), to explore how women reclaimed public spaces over time (Ouaryachi, 2018), and to delve into the political significance of intimacy in gender-separated spaces such as the *hammam* (Graiouid, 2004). Adopting a similar approach, this chapter considers gendered power dynamics through the lens of spatial division. Does gender separation in Sidi Bounouar reflect a form of gender oppression? And how might such an analysis inspire alternative ways of thinking about power when considering gender in a Muslim-majority society?

To explore these questions, relationships among women and men need to be considered in their everyday dimension. To do so, this chapter focuses on foodways as culturally embedded practices encompassing food production, ways of accessing food, distribution, and consumption, examining active fishing spaces and moments of conviviality and food sharing. Food, in fact, serves as a lens to trace power dynamics (Hayes-Conroy and Hayes-Conroy, 2008), value alternative epistemologies (Hammelman et al., 2024), and understand the construction of gender relations and meanings in different contexts (Counihan & Kaplan, 1998). With this in mind, the chapter focuses on foodways, examining active fishing spaces as well as moments of conviviality in cooking and meal sharing. This perspective makes it possible to access the everyday dynamics of gendered separation and interconnection, approaching food as a site of community-making that engages women and men in various ways.

Drawing on this analysis, I argue that gender separation is less sharp than it is often presented in development discourse on gender, and that when separation does occur, it does not necessarily imply unequal power relations; rather, it may reflect an alternative mode of social cohesion. The making of society must be conceived differently in every context, and the "problem of gender equity", together with related articulations of power relations, agency, and justice, identified as central to gender research (Bradshaw et al., 2020, p. 134), should be reframed in accordance with locally grounded understandings of this principle. This paper aims to explore both the interconnections between men and women, demonstrating how these seemingly distinct groups are embedded in continuous everyday relations, and the ways in which they are separated, participating in social life through various roles. The aim is to challenge the assumption that separation inherently implies distance or power inequality—as often suggested by development institutions' discourse on gender as a set of universal skills, attributes, and governmental technologies (Mukhopadhyay, 2014)—and to open space for alternative representations of gender relations and power dynamics in an Amazigh, Muslim-majority society.

2 Alternative Gender Representations: Perspectives from Moroccan Foodways

In this study, "gender" is approached not as a question of identity, but as a set of socially situated roles, practices, and spatial relations. While recognising that gender exceeds a binary framework, the ethnographic fieldwork focused on relations between men and women, reflecting the most salient gendered roles and social expectations encountered in this context, including those related to spatial presence and gendered divisions. The heterogeneity of gendered power dynamics has long been a central concern within non-white feminist scholarship, which has challenged liberal feminist assumptions that equate agency, equality, and emancipation with specific normative forms of visibility and autonomy (Mahmood, 2005). Investigating such heterogeneity requires exploring the interconnections and local meanings of gender role differentiation. These elements are key to constructing alternative representations that challenge universalist and essentialist narratives and to analysing gender outside Western and colonising frameworks, that have structured dominant approaches to gender in many international development programmes, beginning with the Women in Development (WID) paradigm (Moser, 1993).

While patriarchy remains a critical tool for political analysis, relying solely on it to understand gender dynamics can obscure the complexity of local power relations, gender socialisation, and political organisation. Understanding women-men relations benefits from directly examining how power circulates within society and what kinds of interconnections shape gender roles and groups (Ong, 1988). Muslim-majority societies have long been portrayed as inherently marked by unequal power relations, and this orientalist and Islamophobic framing obscures the possibility that such contexts can instead serve as sites for developing alternative perspectives beyond Western/white feminism on women-men relations (Manzoor-Khan, 2022). Enabling this possibility requires attention to the blind spots produced by adopting gender in arbitrary and normative ways, and challenging normalised, implicit assumptions is a key step in this process (Khader, 2017). In line with Zakaria's approach (2021), non-white feminism invites a reading of gender that moves beyond assumptions of inherent conflict and asymmetrical power relations, challenging representations of non-Western women as merely discriminated or segregated subjects and foregrounding their capacity for choice, negotiation, and agency. Kumar (2015) uses "complementarity" as a lens through which to conceptualise equity, valuing the different roles of men and women. Similar efforts can be found in scholarship exploring gender' means in terms of social roles for people involved in development projects in contexts such as Malawi (Riley & Dodson., 2016), India (Bradshaw et al., 2020), and Nigeria, making explicit the "myths" (Cornwall, 2007) that structure development discourse and constrain understandings of women-men relations in non-Western contexts. Moreover, research on Muslim-majority societies has proposed situated approaches to understanding women's political action (Abu-Lughod, 2015; Della Negra, 2024) as well as participation in religious life (Wadud, 1999).

Within this broader critique of reductive gender frameworks, food- and fisheries-related practices offer a particularly revealing yet underexplored site of analysis. As the literature points out (Kleiber et al., 2015), although shellfish gathering is often framed as a women's activity across a range of contexts, including Spain (Frangoudes et al., 2013), Tonga (Malm, 2009), Oman (Bose et al., 2013) and Brazil (Anacleto et al., 2021), the ways in which gender actually shapes this practice *in situ* remains insufficiently examined. Following Kleiber et al. (2015), this study builds on the argument that "a gender approach has considerable potential to improve our understanding of different fishing methods, species caught, and areas fished" (2015, p. 26) while also illuminating gender dynamics themselves. Such an approach offers an entry point into women-men relationships and their power relations from a perspective that has so far received limited attention.

Considering Morocco, some studies use gendered social roles as a lens to move beyond the modernity/tradition duality. Research on sexuality (Glacier, 2017), health (Batnitzky, 2011), healing knowledge (Davis, 2005), and technology (Dodson et al., 2013) offers entry points for accessing different representations of "gender" and "women" in contemporary Morocco. Also the exploration of women's participation in marketplace life highlights the dynamic and polysemic meaning of this category, showing how womanhood conveys "a plurality of coded messages, ideologies, and power relations" (Kapchan, 1996, p. 9). The relation between state and religion has also been examined by analysing the influence of feminism on Moroccan political discourse (Sadiqi, 2024), and by highlighting its contribution to nationalist movements and debates between secular and Islamic feminisms (Mernissi, 2003). Historians have also challenged the orientalist representation of women as passive, invisible "objects" examining Moroccan harems from the fourteenth to the twentieth centuries (Dakhlia, 2024), and proposing interpretations able "to move beyond the simple binary of 'dominant-dominated'" (Hachimi Alaoui & Hibou, 2025, p. 122).

Approaching gender through the production and interpretation of space has offered a means to move beyond binary readings of gender roles. Drawing on bell hooks' work on gendered spaces (2009), which emphasises that power relations must be read through meanings articulated from within social groups and may be obscured or distorted by Western or white epistemological frameworks, Salih (2013) analyses the gender-space relationship in Moroccan society through the lens of transnationalism, focusing on how international mobility reshapes gender roles and how specific "spaces" (such as homes, families, food, and rituals) become sites where womanhood is defined and experienced. This work underlines the need to examine how gender is materially performed within socio-ecological relationships, contributing to the dismantling of binary logics and illuminating the "polyphonic discourse" (*ibid*, p. 94) surrounding gender in non-Western societies.

Moroccan society is marked by cultural and ethnic diversity, reflected, among other things, in foodways. A significant portion of the population identifies as Amazigh (in the Souss Massa region, specifically, as *Tachelhit*), a pre-Arab population that has long faced political oppression (Peyron, 2022). Among Amazigh communities, gender is an entry point principally for highlighting identity-making processes through different forms of artistic expression such as tattoos (Mesouani, 2019), dress and textile practices (Becker, 2006; Dolz, 2019), poetry and music (Hoffman, 2008), often privileging gendered expressions of cultural identity over the relational interconnections between women and men. Within studies of Amazigh populations, these interconnections have been only marginally addressed. While nutritional science explored how dietary changes reflect political, social, and economic transformations (Lotfi et al., 2012), and how state-led agricultural policies, the climate crisis, and the industrialisation of food systems have reshaped Amazigh foodways (Belahsen et al., 2017), these analyses have not been systematically connected to questions of gender roles, except insofar as women are considered primarily in relation to family cooking. Instead, this body of work has tended to focus on structural inequalities affecting Amazigh communities, which are "confronted with issues of food access and security, the highest national poverty" (*ibid,* p.14), or on socioeconomic changes contributing "to the homogenization of dietary models and loss of food diversity" (*ibidem,* p. 16).

In the broader Moroccan context, food and gender have instead been investigated in a more interconnected way. Studies on cooking, healing, and ethnobotanical knowledge (Teixidor-Toneu et al., 2021), as well as analyses of dietary shifts in urban and rural settings due to social and economic change (Zirari, 2020) offer perspectives on changes of gender dynamics in the Moroccan context. Cooking practices are part of this body of work, examining how they are transmitted intergenerationally (Graf, 2024b) and how they are transformed by the use of new technologies by women in the kitchens (Graf, 2024a).

Building on this literature, and in order to explore how spatial gender separation takes place and acquires meaning, this chapter examines how food can act as a medium for understanding relations between women and men. Through this lens, it investigates how gendered food spaces can reveal alternative understandings of women-men separation, moving beyond dominant representations that reduce it to a mere symbol of male oppression, and foregrounding instead the multiple interconnections that unfold between genders through food-related practices and space.

3 Embodied Methods: Sensing and Understanding Gendered Spaces in the field

To delve into the aforementioned dimension, interactions between men and women in Sidi Bounouar are examined through the ways they are enacted in and through food-related spaces. Focusing on this material and relational dimension makes it possible to trace everyday connections and to observe when gendered spatial separations emerge, are negotiated, or become blurred, as well as the power relations they express. This methodological orientation stems from the intention not to "speak for" the people involved, but rather to reflect on what has been directly experienced and co-questioned.

The analysis is grounded in a long-term relationship with Sidi Bounouar's residents. I first met the community in 2015 through a collaboration with an international association of which several inhabitants were members, during research on the food system in the Souss Massa region. Although intermittent, contact continued in the following years through occasional visits linked to other research activities. When

Sidi Bounouar became part of the FoCE project, a basis of trust had already been established with some members of the village. Building on this relationship, four fieldwork periods were carried out in 2024 (April, July, and October) and 2025 (December), each lasting several weeks and drawing on a range of methods.

As a foreigner, I was granted access to both women's and men's spaces, which allowed me to observe how gendered spatial divisions are not fixed but shift according to context, relationships, and moments of collective life. Participant observation, semi-structured and unstructured interviews were combined with photo elicitation, visual methods, focus groups, historical and mental mapping, walking interviews, and participatory digital mapping. Visceral and embodied methods—cooking, eating, harvesting, fishing, playing, and dancing—were central to building relationships and trust, especially with interlocutors with whom I did not share a common language. I speak only a few words of Amazigh (the main language spoken in Sidi Bounouar) and limited Darija (Moroccan Arabic). This linguistic barrier constrained my interactions, but it also afforded people greater control over the information they chose to share, as I could not access their interactions without their consent. Over time, a sense of complicity and a shared hybrid language emerged, as meanings were patiently constructed together.

Food—through its production, processing, and consumption—functioned as a central vector for building connections and for attending to space, movement, and embodied practices, offering a grounded entry point for observing how gender separation is enacted and experienced. It also became a vehicle for building bonds in situations where other methods were not fully usable. Several people in the village, in fact, do not feel comfortable with writing or with more "codified" research methods such as visual tools or interviews. Working through foodways involving fishing, cooking, and eating together made it possible to overcome some of the barriers that created distance and to establish forms of reciprocity. Food also functioned as a space of exchange—of recipes, ingredients, and ideas—and became a medium through which dialogues and methods were co-constructed.

Another important aspect of how this research has been carried out is that gender was not at the centre of my research focus while being in the field. The FoCE project was primarily concerned with understanding the food systems of the Sidi Bounouar community and how their practices of accessing food were changing amid the profound socio-environmental transformations affecting the area. As fieldwork progressed, I came to understand that the marked gender separation I was witnessing could not be interpreted through the analytical frameworks commonly employed in development projects informed by WID approaches, in which gender separation is often framed as a direct reflection of men's domination of women. Instead, these observations challenged my own analytical categories and invited me to imagine alternative meanings for this spatialised and highly visible practice.

Building on this material dimension of everyday life, foodways and fishing spaces will be central to examining how gendered divisions and interconnections take concrete form and how they articulate power dynamics.

4 Observing Gendered Spaces in Everyday Life

From food production to food consumption, different conturations of gendered separation can be observed. The analysis reveals gender-related spatialities across two main sites: the fishing spaces (shores, boats, and caves) and the spaces of food transformation and consumption (homes and ceremonial sites).

4.1 The Fishing Spaces: Shores, Boats, and Caves

Fishing spaces are shared sites where specialised activities unfold, often simultaneously. Humans and non-humans, as well as women and men, encounter one another on the intertidal beach, following the periodic rhythm of the tide. Ecological changes have buried under sand several rocky areas that once hosted forms of biodiversity of high value for fishers, and constituted key fishing grounds. As a result, only limited rocky areas remain exposed and accessible, and these are now subject to intense overexploitation, making it increasingly difficult to respect ecological rhythms. Fishers are commonly distinguished between those recognised as more experienced, whose practices remain ecologically attuned and sustain fishing through selective harvesting, and newcomers, who have turned to the sea after losing previous employment and tend to adopt a more functionalist and extractive approach. Artisanal boats, anglers, (illegal) net-fishers, octopus hunters, and shellfish harvesters all gather where the

Fig. 1 The intertidal beach of Talat Bouazaa (Author's photo, July 2024)

receding sea exposes rocks (Fig. 1) covered with mussels and other shellfish such as sea snails, *Arca noae*, gooseneck barnacles, sea urchins, razor shells, and crabs (Fig. 2).

Women usually go in pairs or in small groups to nearby beaches on foot or with donkeys to carry their catch, or to more distant shores reached by bus. Groups form along lines of friendship and/or village belonging. Harvesting mussels and other shellfish can be performed in different ways, and the technique used defines divisions among women. Some use the *Askarkaz* (a metal hook used to pry mussels off the rocks) indiscriminately, like a rake, while others are more selective, targeting only adult, healthy mussels. This technical difference reflects divergent modes of relating to the environment and to sea life, revealing distinct situated ethics within the practice of gathering and creating internal divisions among women working in the same spaces. Within each group and despite the increasing scarcity of shellfish, gathering remains a collective activity: women share shellfish, point out better spots, and move together across the rocks. Harvesting shellfish is physically demanding and dangerous: the wind is strong, the rocks are slippery, and the rising tide quickly makes the terrain hazardous. While people navigate the rocks skilfully, the risks are ever-present. Women, wearing multiple layers of clothing and veils, arrive together and move as a group, occasionally splitting into sub-groups while remaining within visual proximity. Encounters

Fig. 2 Marine products harvested along the coast (Author's photo, July, 2024)

with women from other villages are common, but exchanges are brief and pauses are rare.

Men also go to these rocks. Those who gather shellfish or hunt octopus (a practice occasionally carried out by women as well) move alone. Their primary social identity remains that of "fishers", and while harvesting, they also collect sea potatoes or worms for bait. When catching octopus, they may share portions with anglers, as it is considered excellent bait for seabream. Fishers targeting finfish generally operate in groups. Artisanal fishing takes place on boats with up to five people, staying close to the shore. Angling can be done individually, but most often occurs collectively. Fishermen form a cohesive social group, supporting one another even across village boundaries—offering rides or sharing surplus catch. They commonly refer to others as "one of the fishers", expressing a shared identity. In fact, specialisation among the men does not strongly define belonging. Many alternate between shellfish gathering, octopus hunting, rod fishing, and occasionally artisanal fishing; those who practice only one technique—which is common—are still considered part of the group. Expert shellfish gatherers are called *raiss*, the same term used for the shipowners and coordinators of artisanal fishing. Even if the formalisation of artisanal fishing and the decline of marine life affect this community, their unity persists in everyday relationships. Men also use other areas of the rocky coastline, "the caves", where they have built, over time, shelters or hollows carved into the rocks facing the sea. These structures serve both as gear storage and as temporary homes where men spend time, from a few days to several months, living exposed to the ocean and close to colleagues and friends (Fig. 3). Although the state has attempted to demolish the caves as part of coastal land-use planning, these spaces continue to be used; where this is no longer possible, fishers gather instead in the new formal fishing complexes built by the public sector for artisanal fishing—spaces that, once again, host mainly men (Fig. 4). Together, these spaces function as spaces of their own, through which the male community of fishers' cohesion is articulated, enacted, and continually reinforced.

Fig. 3 Issoh's caves, and interior of one cave (Author's photo, July 2024, December 2025)

Fig. 4 Formal fishing complex (Départment de la Peche Artisanale) of Aglou (Ginevra Montefusco, April, 2024)

4.2 Homes and Ceremonial Spaces

While caves and boats are the only fishing places where clear gendered spatial separation occurs, divisions between men and women also emerge in food processing and consumption. When men are alone, they cook; but when women are present, they take charge of meal preparation. The kitchen is a central gathering space for women. Although it may be tempting to idealise this environment, dynamics are not always harmonious: moods shift from day to day, and once the cooking skills of a woman from outside the household are recognised, the kitchen is often readily entrusted to her, thus suggesting not only a certain porosity in access to this familiar space, but also an ambivalent relationship to the everyday responsibility of cooking for the family.

Participant observation suggests that the sharing of everyday meals varies across families. In some households, men and women eat together; in others, particularly during large gatherings, meals are gender-separated, with women eating in a different space or waiting until the men have finished. This arrangement allows women to maintain their own social sphere, often characterised by laughter, singing, and animated conversation. When asked, they expressed no desire to join the men's table, which is typically quieter and more restrained. After meals, women sometimes sing or dance, either in celebration (e.g., for rainfall or visitor's arrival or departure) or simply for enjoyment. Although men may join, participation is selective: shared celebrations occur only when men are welcomed by the women. If unease arises women stop from performing, and the men leave almost immediately.

Gendered spatial separation becomes particularly visible during extraordinary events. Women, unless they choose not to participate, actively take part in public village gatherings and are often central to their organisation. The coexistence of gender separation and communal belonging becomes apparent during public music performances, where people spontaneously divide into groups of men, young men, and women. Although seated separately, they attend the same event and share the same music, participating together in a collective moment.

Another example of the coexistence of gender separation and communal participation is the Tigri Fish Festival, organised since 2015 by the Amoud Association (sometimes in partnership with Slow Food) to promote local fishing practices and problematic issues. Women's participation is primarily articulated through cooking, with the only public moment specifically dedicated to them taking the form of a cooking competition, with preparations beginning days in advance. For the remainder of the festival, most village women spend considerable time in the association's kitchen where, dressed much as they would be at home, they cook together, coordinate tasks informally, and create a lively atmosphere of conversation, song, and laughter. While men may enter the space, access is regulated through informal social cues, reflecting women's control over the space. Brief visits by outsiders, such as local dignitaries wishing to greet the women, are permitted but carefully managed. Notably, only one man was regularly present in the kitchen: a former chef with expertise in preparing *tajines*. During the festival, men from the Amoud Association often encouraged women to eat in the local restaurant, to make them visible in public spaces, in line with institutional and foreign expectations. These attempts were eventually abandoned, as women preferred to eat together in the kitchen. When asked why, their responses emphasised sociality and choice, rather than exclusion:

"my friends are here, I don't want to go there"; "I'm not interested"; "I prefer it here". A similar dynamic emerged during public institutional events— such as speeches or formal discussions involving local authorities and international NGOs— where festival organisers insisted on women's participation. For many women, however, these occasions were dismissed as irrelevant, described as "just *bla-bla*", highlighting a disconnect between externally valued forms of public participation and their meaningful modes of engagement.

Celebrations see women and men participating in parallel groups and can take place in private homes or in shared spaces such as the headquarters of the Tigri Cooperative, the Amoud Association, or the "House of the Mosque"—named in Amazigh *tikmil jikh* ("the people's house"), a space behind the village mosque with a kitchen and several communal rooms. Some events are exclusively for women, while others are structured in two parts, enabling separate gatherings. In these cases, men eat first, usually in relative silence, followed by women, who gather elsewhere or later, eating, singing, dancing, and socialising late into the night. Sometimes these events rely on women hired to prepare food for all attendees.

Men are entirely excluded from women-only events, such as *marida*—a recovery gathering in which women visit a sick member of their circle to eat, sing, and pray for her healing. These invitation-only gatherings often last until late at night and reveal the internal dynamics of the women's community. Friendships and kinship ties shape smaller subgroups, sometimes excluding others. While these distinctions are less visible in collective events, they nonetheless influence the village's social fabric, forming groups shaped more by personal bonds than by a shared gender identity.

Overall, these domestic and ceremonial spaces point to a gendered organisation of food-related practices that is fluid and context-dependent rather than fixed or hierarchical. Moments of separation do not disrupt collective life but instead sustain parallel social spheres through which belonging and community cohesion are produced and maintained.

5 Gender Separation and/as Community-making

As discussed throughout the chapter, gender separation is very common in everyday food-related activities in Sidi Bounouar. Yet this does not mean that spaces and practices cannot, at times, be shared or become common. The specialisations and spatial dimensions embedded in foodways and fishing practices help illuminate how separation and sharedness are intertwined, and how division can both coexist with and actively produce forms of togetherness, community-making, and belonging. When gender separation occurs, it does not imply social "backwardness", or hierarchical power relationships between men and women. Rather, spatial separation constitutes a situated social configuration through which specific forms of women's agency are produced, intelligible only when adopting a locally grounded perspective on gender dynamics. Rather than reflecting imposed patriarchal segregation, women's presiding over distinct spaces is often intentional and strategic, enabling forms of belonging, trust, care, and relationality, in resonance with bell hooks' reflections on gendered spaces as worlds of belonging (2009). This counter-narrative challenges white and Western assumptions that equate emancipation exclusively with a single model of liberation that includes public visibility in mixed-gender spaces (Mohanty, 1984). Building on this perspective, closer engagement with everyday women-men relations allows gender to be approached as neither rigid nor absolute, calling for a shift away from interpretative frameworks commonly adopted in WID perspective and white feminism approaches when addressing Muslim-majority societies.

The form of separation observed in Sidi Bounouar reflects an embodied experience of gender roles and relations, sharply distinguishing men's and women's everyday lives by clearly defining when spaces can be shared and when they remain separated. In this context, gender roles are lived and enacted rather than explicitly verbalised and do not circulate as part of collective discourse. Criticism directed at men tends to focus on individual behaviour rather than being framed in relation to patriarchy. Yet gender is mobilised strategically in communication with outsiders, including myself. For example, F. (one of the women I was closest to) referred to her circle as "*les filles*" when inviting me to join them, using a gendered French label even though such terminology is absent within the group itself, which is experienced less as a gender-based network than as a circle of friends shaped by individual relationships and everyday affinities.

The fact that this category is mobilised primarily in relation to an external gaze does not mean that gender division is not actively enacted in everyday interactions. Women are absent from certain fishing spaces and practices, and men are excluded from women's spaces during ceremonies and in daily life, but these restrictions are not explicit social prohibitions; rather, they operate as embodied habits that do not prevent actual interactions, nor do they constrain the various forms of interconnection, through knowledge circulation and shared practices, that take place among them (Gururani, 2002). Intra-group solidarity is strong, as fishing practices are not only dangerous but also require collaborative work. Most men and women cannot swim, yet they confront the ocean daily in its intertidal spaces, which demands collective attentiveness to the sea's rhythms and movements. As an experienced fisher put it, "fishing is a collective art", one that relies on community support and requires shared effort across all stages of the work. Gathering, in particular, depends on collective awareness, while angling and artisanal fishing depend on community cooperation to physically capture larger prey. These latter two fishing styles, together forming a single fishing community in practice, are interconnected through forms of reciprocal, long-distance support. One example is when shore-based fishers warn those at sea that the waters are becoming rough by lighting a fire whose smoke can be seen from offshore.

As this example suggests, everyday community-making practices, both within gendered groups and across them, often unfold through non-human beings. The sea and sea life act as crucial intermediaries in human relationships, shaping interactions both inside and between men's and women's groups. Fish, mussels, and other seafood are given as gifts: wishes for recovery, hopes for a good catch, or gestures of friendship when someone's harvest is poor. These exchanges reveal how non-human beings are embedded in social life and in the processes of community-making. They are also central to self-identity: being a "fisher" is a key way through which many—especially men—define themselves. The interaction of multiple species in the coastal space (mussels, sea snails, crabs, octopuses, fish) shapes human relations and creates interspecies "webs of social relations" (Tsing, 2014, p. 28) that affect relationships, senses of belonging, and identity. Non-human beings form a crucial part of the community's sense of kinship (Haraway, 2016) and become important intermediaries in both intra- and inter-gender connections.

Food is the main vector of women-men interconnections during ceremonies and social rites. Even when groups celebrate in spatially separated settings, this division does not imply disconnection; what might appear as "segregation" operates, in practice, as a specific form of social cohesion. As in festivals and musical performances, where the village's social fabric participates in the same event from separate clusters, the separation between men and women reflects a way of being a community made up of distinct yet interconnected groups, rather than one governed by a rigid divide. Women's spaces, in fact, are porous, and selected men may enter and participate when their presence is collectively welcomed. The chef preparing *tajines* during the festival is one example among several occasions in which I observed men singing, playing music, or engaging in conversation with women during gatherings otherwise considered "women's" spaces. Gender separation is context-dependent, shaped by relationships, mood, and individual agency; it is dynamic and adapted to people and desires. Such flexibility highlights that women exercise control over their spaces, determining who can enter and who remains outside.

But interconnections do not exist only when they are visible. Spatial separation, in fact, can become a lens through which to understand women-men relations from a different analytical angle. While in urban contexts the role of the kitchen as a primarily women's space is changing (Zirari, 2020), in Sidi Bounouar it retains its centrality. The circulation of power between men and women, which allows women to maintain their own separate spaces in Muslim-majority societies (Sawalha, 2014) is a crucial element for understanding how such spaces can be places of comfort, intimacy, and participation (Hebbouch, 2019). These meanings are key for interpreting the gender separation experienced in Sidi Bounouar and for understanding why, even when invited into public spaces, women in the village preferred to remain at home or in the kitchen. For them, substantive participation in social life does not necessarily unfold through public visibility; rather, it relies on the private dimension, where relations, responsibilities, and forms of belonging are cultivated.

A symbolic yet revealing manifestation of the ongoing interplay between the private sphere and the sense of community belonging emerges in the lyrics of the songs performed by women during ceremonies and social gatherings. As recorded in my fieldnotes:

> The songs are repetitive, built around a single phrase, sung in variations of rhythm and tone. Each last around 10–15 minutes and centres on themes of community and kinship. For example, they say "What a happy day, we are all together and I am with my family"; "My brother is always at my right, my cousins always on my left"; or "It is so beautiful to sing together on this day of celebration". The sound emerges gradually, like a

wave, initiated by one or two voices, growing into a collective peak, then fading into scattered words. After a pause filled with chatting, another song begins.

The importance of belonging that transpires in these lyrics permeates everyday life despite the spatial separation of genders. Rather than signalling submission, spatial and practice-based distinctions are forms of "being together", shaped by social norms and shared expectations. Gender distinctions may reflect a broader patriarchal order, as it may appear in other situations, without necessarily translating into relations of subordination or unidirectional power.

6 Conclusion

The chapter offers one possible way of unsettling hegemonic discourses on gender, and more specifically on gender equity as a self-evident analytical category. By foregrounding food as a primary space of community-making, the case of Sidi Bounouar serves as a reminder that gender should never be taken for granted: each social context produces its own configurations of gender relations and elaborates, in contingent ways, what "community" means.

Through the analysis of foodways and fishing spaces in Sidi Bounouar, this chapter has proposed an interpretation of gendered spatial separation that differs from those often adopted in development discourse and white feminist perspectives, where women-men division is commonly assumed to reveal power inequality and male domination. By grounding the analysis in the spatial forms through which relations between men and women take shape, the chapter has examined interconnections, divisions, and their social meanings, questioning their implications for dominant politics of visibility that are often misread as direct indicators of power relations. It has argued that the gendered division through which Amazigh and Muslim-majority societies are often interpreted, particularly within development and political discourses, is complex, dynamic, and porous, rather than fixed and rigid. Interconnections frequently occur—whether through shared spaces and reciprocity, particularly when food and non-human beings are involved, or through the selective presence of men in women's spaces. Even when gendered separations shape everyday practices, these do not inherently imply domination; rather, they may reveal alternative modes of community-making that diverge from Western institutional models, in which gender equality is expected to materialise through identical forms of participation by men and women. In this context, power circulates between men and women, making private spaces sites of intimacy and negotiation, where boundaries are often defined and regulated by women.

This reading also speaks to broader critiques of development discourse, which over its long trajectory of adopting gender as an analytical framework has tended to oversimplify it by codifying and reducing the vastly different experiences of women across contexts. In Sidi Bounouar, residents themselves mobilise the category of gender when presenting their community to external actors, drawing on it strategically to legitimise projects in the eyes of funding institutions. Gender thus operates simultaneously as an externally imposed framework and as a locally reworked resource, revealing the ambivalent ways in which development categories are appropriated, negotiated, and translated within everyday life.

Bibliography

Abu-Lughod, L. (2015) *Do Muslim Women Need Saving?* First Harvard University Press paperpack edition. Harvard University Press.

Adolfsson, J. S. & Madsen, O. J. (2020). 'Nowadays There Is *Gender*': 'Doing' Global Gender Equality in Rural Malawi. *Theory & Psychology, 30*(1), 56–76. https://doi.org/10.1177/0959354319879507

Adusei-Asante, K., Hancock, P., & Oliveira, M. (2015). Gender Mainstreaming and Women's Roles in Development Projects: A Research Case Study from Ghana. In V. Demos & M. Texler Segal (Eds.), *Advances in Gender Research* (pp. 175–198). Emerald Group Publishing Limited. https://doi.org/10.1108/S1529-212620150000020019

Ajaz, N. (2022). Understanding Gendered Power Relations Thorugh Transect Walk and Spatial Mapping. *Pakistan Journal of Social Research, 04*(03), 372–379. https://doi.org/10.52567/pjsr.v4i03.726

Anacleto, A., Oliveira, A. M. P. D., Maia, A. P., Rodrigues, K. A. D. C., & Silva, T. M. D. (2021). Women Shellfish Collectors: Economic and Social Implications of the Shellfish Extractivism Perna perna (Linnaeus) in Paraná Coast Brazil. *International Journal of Business Administration, 12*(4), 34. https://doi.org/10.5430/ijba.v12n4p34

Bagheri, N. (2014). Mapping Women in Tehran's Public Spaces: A Geo-Visualization Perspective. *Gender, Place & Culture, 21*(10), 1285–1301. https://doi.org/10.1080/0966369X.2013.817972

Batnitzky, A. K. (2011). Cultural Constructions of 'Obesity': Understanding Body Size, Social Class and Gender in Morocco. *Health & Place, 17*(1), 345–352. https://doi.org/10.1016/j.healthplace.2010.11.012

Becker I. C. (2006) *Amazigh Arts in Morocco: Women Shaping Berber Identity.* Springer (Lecture notes in computer science, 4117).

Belahsen, R., Naciri, K., & El Ibrahimi, A. (2017). Food Security and Women's Roles in Moroccan Berber (Amazigh) Society Today. *Maternal & Child Nutrition, 13*(S3), e12562. https://doi.org/10.1111/mcn.12562

bell hooks. (2009). *Belonging: A Culture of Place.* Routledge.

Bose, S., Al-Kindy, F., Al-Balushi, A., & Rajab, M. M. (2013). Accounting the Unaccounted: Case of Women's Participation in Shellfish Harvesting in the Sultanate of Oman. *Gender, Technology and Development, 17*(1), 31–53. https://doi.org/10.1177/0971852412472123

Bradshaw, S., Linneker, B., & Sanders-McDonagh, E. (2020). It's Gender Jim, But Not As We Know It … A Critical Review of Constructions of Gendered Knowledge of the Global South. *European Journal of Women's Studies, 27*(2), 128–144. https://doi.org/10.1177/1350506819847233

Cornwall, A. (2007). Myths To Live By? Female Solidarity and Female Autonomy Reconsidered. *Development and Change, 38*(1), 149–168. https://doi.org/10.1111/j.1467-7660.2007.00407.x

Counihan, C., & Kaplan, S. L. (1998). *Food and Gender. Identity and Power.* Taylor and Francis (Food in History and Culture, v. 1).

Dakhlia, J. (2024) *Harems et sultans: genre et despotisme au Maroc et ailleurs, XIVe-XXe siècle.* Anacharsis (Collection Essais).

Davis, D. K. (2005). Indigenous knowledge and the desertification debate: problematising expert knowledge in North Africa. *Geoforum, 36*(4), 509–524. https://doi.org/10.1016/j.geoforum.2004.08.003

Della Negra, C. (2024). *Questa terra è donna. Movimenti femminili e femministi palestinesi.* AstArte.

Dodson, L. L., Sterling, S. R., & Bennett, J. K. (2013). Minding the Gaps: Cultural, Technical and Gender-Based Barriers to Mobile Use in Oral-Language Berber communities in Morocco. In *Proceedings of the Sixth International Conference on Information and Communication Technologies and Development: Full Papers—Volume 1. ICTD 2013: International Conference on Information and Communication Technologies and Development* (pp. 79–88). ACM. https://doi.org/10.1145/2516604.2516626

Dolz, S. (2019). Fish in the Desert—North Africa's Textile Tradition Between Indigenous Identity and Exogenous Shifts in Meaning. *Textile Society of America Symposium Proceedings* [Preprint]. https://doi.org/10.32873/unl.dc.tsasp.0027

Fondazione Slow Food. (2025). Cozza Wakkad Tigri—Presìdi Slow Food. https://www.fondazioneslowfood.com/it/presidi-slow-food/cozza-wakkad-tigri/. Accessed on November 28, 2025.

Frangoudes, K., Marugán-Pintos, B., & Pascual-Fernandez, J. J. (2013). Gender in Galician Shell-Fisheries: Transforming for Governability. In M. Bavinck et al. (Eds.), *Governability of Fisheries and Aquaculture* (pp. 241–261). Springer Netherlands (MARE Publication Series). https://doi.org/10.1007/978-94-007-6107-0_13.

Glacier, O. (2017) *Femininity, Masculinity, and Sexuality in Morocco and Hollywood*. Springer International Publishing. https://doi.org/10.1007/978-3-319-53285-1

Graf, K. (2024a). Cooking With(Out) Others? Changing Kitchen Technologies and Family Values in Marrakech. *The Journal of North African Studies, 29*(4), 575–600. https://doi.org/10.1080/13629387.2022.2056448

Graf, K. (2024b). *Food and Families in the Making* (1st ed). s.n. (Food, Nutrition, and Culture, 8).

Graiouid, S. (2004). Communication and the Social Production of Space: The *hammam*, the public sphere and Moroccan women. *The Journal of North African Studies, 9*(1), 104–130. https://doi.org/10.1080/1362938042000292324

Gururani, S. (2002). Construction of Third World Women's Knowledge in the Development Discourse. *International Social Science Journal, 54*(173), 313–323. https://doi.org/10.1111/1468-2451.00384.

Hachimi Alaoui, N., & Hibou, B. (2025). Les enjeux de genre dans les sociétés islamiques. Entretien avec Jocelyne Dakhlia. *Sociétés politiques comparées. Revue européenne d'analyse des sociétés politiques* [Preprint]. https://doi.org/10.36253/spc-17459

Hammelman, C., Levkoe, C. Z., & Reynolds, K. (2024). *Radical Food Geographies: Power, Knowledge and Resistance*. Bristol University Press (Food and society).

Haraway, D. (2016). *Staying with the Trouble: Making Kin in the Chthulucene*. Duke University Press (Experimental futures. Technological lives, scientific arts, anthropological voices).

Hayes-Conroy, A., & Hayes-Conroy, J. (2008). Taking Back Taste: Feminism, Food and Visceral Politics. *Gender, Place & Culture, 15*(5), 461–473. https://doi.org/10.1080/09663690802300803

Hebbouch, S. (2019). Crafting Sociability in Female Spiritual Practices: The Case of Boutchichiyyat. *European Journal of Multidisciplinary Studies, 4*(2), 121. https://doi.org/10.26417/ejms-2019.v4i2-545

Hoffman, K. E. (2008). *We Share Walls: Language, Land, and Gender in Berber Morocco*. Blackwell Pub (Blackwell studies in discourse and culture, 2). https://doi.org/10.1002/9780470694145

Kapchan, D. A. (1996). *Gender on the Market: Moroccan Women and the Revoicing of Tradition*. University of Pennsylvania Press (University of Pennsylvania Press new cultural studies).

Khader, S. J. (2017). Transnational Feminisms, Nonideal Theory, and 'Other' Women's Power. *Feminist Philosophy Quarterly, 3*(1). https://doi.org/10.5206/fpq/2016.3.1

Kleiber, D., Harris, L. M., & Vincent, A. C. J. (2015). Gender and Small-Scale Fisheries: A Case for Counting Women and Beyond. *Fish and Fisheries, 16*(4), 547–562. https://doi.org/10.1111/faf.12075

Klenk, R. M. (2004). 'Who is the Developed Woman?': Women as a Category of Development Discourse, Kumaon, India. *Development and Change, 35*(1), 57–78. https://doi.org/10.1111/j.1467-7660.2004.00342.x

Kumar, A. (2015). *Complementing Gender Analysis Methods*. Social Science Research Network. https://papers.ssrn.com/abstract=2604104. Accessed on November 18, 2025.

Lotfi, R., Loukid, M., & Hilali, M. K. (2012). Nutritional Status of Menopausal Women in the Rural Area of Marrakech (Morocco). *International Journal of Science and Engineering Investigations, 1*(1), 104–108.

Mahmood, S. (2005). *Politics of Piety: The Islamic Revival and the Feminist Subject*. Princeton University Press.

Malm, T. (2009). Women of the Coral Gardens: The Significance of Marine Gathering in Tonga. *Traditional Marine Resource Management and Knowledge Information Bulletin, 25*, 2–15.

Manzoor-Khan, S. (2022). *Tangled in Terror: Uprooting Islamophobia*. Pluto Press.

Mernissi, F. (2003). *Beyond the Veil: Male-Female Dynamics in a Modern MUSLIM Society* (Rev). Saqi Books.

Mesouani, H. (2019). *Inked Bodies, Blank Pages; A Study of Amazigh Tattooing*. Illinois State University. https://doi.org/10.30707/ETD2019.Mesouani.H

Mohanty, C. T. (1984). Under Westerns Eyes: Feminist Scholarship and Colonial Discourses. *Boundary, 2*(12), 333–358. https://doi.org/10.2307/302821

Moser, C. (1993). *Gender Planning and Development: Theory, Practice and Training*. Routledge.

Mukhopadhyay, M. (2014). Mainstreamin Gender or Reconstructing the Mainstream? Gender knowledge in development. *Journal of International Development, 26*(3), 356–367. https://doi.org/10.1002/jid.2946.

Murdoch, J. (Ed.). (2006). *Post-Structuralist Geography: A Guide to Relational Space*. SAGE.

Newcomb, R. (2006). Gendering the City, Gendering the Nation: Contesting Urban Space in Fes, Morocco. *City & Society, 18*(2), 288–311. https://doi.org/10.1525/city.2006.18.2.288

Ong, A. (1988). Colonialism and Modernity: Feminist Re-presentations of Women in Non-Western Societies, 3(4), 79–93.

Ouaryachi, F. (2018). *Public Space in Morocco; a History of Gender Segregation.* http://hdl.handle.net/11089/39880

Peyron, M. (2022). *The Berbers of Morocco: A History of Resistance* (Paperback). I.B. Tauris.

Riley, L., & Dodson, B. (2016). 'Gender Hates Men': Untangling Gender and Development Discourses in Food Security Fieldwork in Urban Malawi. *Gender, Place & Culture, 23*(7), 1047–1060. https://doi.org/10.1080/0966369X.2015.1090407

Robson, E. (2000). Wife Seclusion and the Spatial Praxis of Gender Ideology in Nigerian Hausaland. *Gender, Place & Culture, 7*(2), 179–199. https://doi.org/10.1080/713668870

Rodriguez, C. (2001). Shattering Butterflies and Amazons: Symbolic Constructions of Women in Colombian Development Discourse. *Communication Theory, 11*(4), 472–494. https://doi.org/10.1111/j.1468-2885.2001.tb00254.x

Sadiqi, F. (2024). *Women in Morocco.* Oxford University Press.

Salih, R. (2013). *Gender in Transnationalism.* Routledge. https://doi.org/10.4324/9780203383148

Samarasinghe, V. (2014). Lost in Translation? Keeping Women's Issues Alive in Gender and Development Initiatives. *Development in Practice, 24*(1), 30–38. https://doi.org/10.1080/09614524.2014.867303

Sawalha, A. (2014). Gendered Space and Middle East Studies. *International Journal of Middle East Studies, 46*(1), 166–168. https://doi.org/10.1017/S0020743813001359

Slow Food. (2025). *Presidia.* https://www.slowfood.com/biodiversity-programs/presidia/. Accessed on November 28, 2025.

Teixidor-Toneu, I., Elgadi, S., Zine, H., Manzanilla, V., Ouhammou, A., & D'Ambrosio, U. (2021). Medicines in the Kitchen: Gender Roles Shape Ethnobotanical Knowledge in Marrakshi Households. *Foods, 10*(10), 2332. https://doi.org/10.3390/foods10102332

Terras, E. (2024). Taghart. *Network Siyada.* https://en.siyada.org/elias/videos-and-images/new-documentary-taghart/. Accessed on November 28, 2025.

Tsing, A. (2014). More-than-Human Sociality. A Call for Critical Description. In *Anthropology and Nature* (pp. 38–54). Routledge.

Wadud, A. (1999). *Qur'an and Woman: Rereading the Sacred Text from a Woman's Perspective* (2nd ed.). Oxford University Press.

Zakaria, R. (2021). *Against White Feminism.* Hamish Hamilton.

Zirari, H. (2020). Getting Out of the Kitchen! Reshaping Gender Relations and Food Practices in Casablanca. In *Eating in the City. Socio-Anthropological Perspectives from Africa, Latin America and Asia.* Quae.

CHAPTER 6

A Food Festival as a Site of Research Restitution: Reflections on Relationality and Legitimation

Paola Minoia, Beatrice Ferlaino, Ginevra Montefusco, and Tamara Taher

Abstract This chapter examines power dynamics and relationality in field research conducted by a group of women researchers involved in a project on foodways in the Moroccan region of Souss Massa. Foodways are understood as traditional and ritual food-related practices encompassing production, preparation, sharing, and consumption in everyday and festive contexts. The research is based on participant observation and visual methods carried out during a fish festival in the "micro-village" of Sidi Bounouar, a Slow Fish presidium for wakkad mussels, hand-harvested and traditionally smoked outdoors. The paper explores

P. Minoia (✉) · B. Ferlaino · T. Taher
Department of Cultures, Politics and Society, University of Turin, Turin, Italy
e-mail: paola.minoia@unito.it

G. Montefusco
Department of Historical and Geographical Sciences and the Ancient World (DiSSGeA), University of Padua, Padua, Italy

P. Minoia (ed.), *Foodways As Sites of Knowledge and Transformation*, Emplacing Food, https://doi.org/10.1007/978-3-032-28126-5_6

the challenges and opportunities of qualitative research within a triangulated field involving local community groups, particularly women's and fishers' associations, local institutional representatives and foreign Slow Food delegates supporting sustainable food economies and local tourism potential, and the researchers themselves, engaged in feminist, solidarity-based research aimed at decolonizing academia. Through a reflexive approach, the paper deconstructs fieldwork power dynamics and processes of legitimation that can enable reciprocity while also producing misunderstandings or unexpected interactions.

Keywords Moroccan foodways · Coastal and ocean livelihoods · Food festival · Participatory visual research · Feminist discomfort

1 Introduction: The Local and International Dimensions of Sidi Bounouar's Festival

It is a sunny morning in Sidi Bounouar. The first thing we see when we step out of the house we are renting is the immensity of the Atlantic Ocean, which, with its gusts, makes the warm October sun of Southern Morocco feel colder. We turn right and greet the same group of men who lean against the mosque's wall every day. In less than five minutes, we arrive at the headquarters of the Amoud Association. Today is the third and last day of the festival, and everything works smoothly. The three traditional ovens, made of clay and placed at the entrance of the headquarters for the occasion, are already baking bread. A., the appointed official waiter of the entire festival, and dressed elegantly in a self-made hat reminiscent of those worn by chefs and two overlapping waistcoats under an oversized suit, is running up and down the stairs that connect the kitchen with the terrace, bringing tea, coffee, bread, oil, and amlou for breakfast to the guests of the festival. M., wearing a yellow, fluorescent vest, is collecting the trash thrown by the people arranging their stands. Inside the kitchen, the women of the village cooperative are talking and singing, cleaning dried fava beans and peas to cook the b'ssara soup for lunch. On the terrace, groups of people are having breakfast. Around one table, some fishermen from the local association sit together, having set up their stand in the yard to sell tajines of goose barnacles and other seafood. At another table, young soldiers in uniform share a meal. A third table is occupied by a group of vendors sipping tea before setting up their stands. A fourth table hosts foreign guests from Canada, South Korea, and Italy, including us. They—and we—are all involved in the international session scheduled for midday. H., the president of the Amoud Association, is nervous. This is the most important institutional moment of the entire festival, and he wants it to be an occasion to present all the networking efforts he has made over the past year to build solid relationships with international groups and national institutions. It will take place in a traditional Amazigh tent, but it will end up being—as H. wrote to us via sms—"a disaster".

The event starts off on the wrong foot. At the scheduled opening time, only men and international guests are inside the tent, while none of the women of the village are present. This absence resonates with an episode of days earlier, when we screened photos and a documentary inside the fishermen's association, which the community enjoyed, especially seeing one elder featured with his self-organized ocean museum. The women's hands too appeared in our visuals, portrayed while harvesting seashells, cooking, and singing together. When some of them hesitantly took the front-row seats, local TV operators began filming their faces. Visibly upset, several withdrew under their Addal veils and left the front rows. So, to ensure the presence of some women at the round table, the session chair gestures for us to call the most active among them and have them sit in the front rows. At last, after a delay of an hour and a half, the meeting can begin. Though we are not seated at the speakers' table, the local fishermen are expecting to hear and discuss the research we have conducted together over the past few weeks. Actually, we had planned to present it to the villagers the day before, but as the women were busy with the tajine contest, attendance would have been too low. The organizer thus suggested postponing it to this policy roundtable.

It soon becomes evident that H. wants to prioritize the foreign guests seated at the speakers' table. Translation is provided by a single person moving between Darija (Moroccan Arabic), Amazigh, French and English. The first to speak is K., from South Korea, who presents the activities of the Korean Slow Fish community. He is followed by an Italian guest who has previously collaborated with H., and who delivers a speech in French on sustainable tourism as a means of valorizing local heritage. Next, a representative of a national Moroccan institution outlines general principles of blue economy solutions for coastal communities. H. then attempts to bring in another expert from Italy via an online connection, even though he is not listed in the programme. After several failed attempts, H. eventually gives up. The floor is then given to J. from Canada, who speaks about the role of local communities in protecting the marine environment. Tensions begin to surface when J.'s slides display a map depicting a border separating Western Sahara from Morocco. Members of the audience voice their objections, and J. acknowledges his lack of awareness of

the political implications of the map. In response, H. intervenes to reaffirm the importance of recognizing Morocco's sovereignty over the Sahara region. Someone from the audience criticizes the overrepresentation of foreign speakers who appear disconnected from the local context.
After three hours of presentations in multiple languages, exhaustion is setting in, but the meeting continues with a contribution by the president of a fishing association, who speaks in Darija about the importance of preserving the rhythms of fishing, rooted in local traditions and understood as essential to the protection of the ecosystem. His words stand in sharp contrast to the more standardized proposals put forward by international and institutional speakers, who have largely focused on externally driven solutions and on the centrality of tourism as a central development strategy.
We wait until this point to finally present our findings. Since we are not included in the official programme, we hope to at least fit into the Q&A session. However, given the significant delay and the unexpected direction of the panel discussion, it has become clear that there is no longer time for us to present our work. Despite H.'s objections— he avoids making eye contact with us— during the Q&A we take the stage and hurriedly share our findings. We present our community map, highlighting local knowledge of the sea and fish, and showing how the ecosystem is affected by regional planning and upstream infrastructural changes. Our presentation opens space for the fishermen to speak as well. Finally, with an enormous delay, at 2:30 p.m., a tribute is held for a young fisherman, well-known and beloved by the local community. He is the first among them to earn a Ph.D. while continuing to work as a traditional fisherman.
A picture is taken, and people slowly make their way toward the terrace, where lunch—prepared by many of the village women—is ready to be served. [notes from the field, 10/26/2024, Sidi Bounouar].

This vignette recounts a situation typical of fieldwork, in which activities, despite being jointly planned with local actors, do not unfold according to schedule, but are instead overtaken by emergent events and shifting priorities. We focus, in particular, on a crucial moment in the participatory research process, that of reporting back. In this case, this was intended to take place during a festival, an event expected to ensure a wide audience, but also one that concentrates organizational pressures and complex political relations among multiple local actors and outsiders.

Our research was based in Sidi Bounouar, a tiny village on the Moroccan Atlantic coast, where 200 people live harvesting intertidal shellfish, hand-catching octopus, and practicing rod fishing. These food practices are referred to as *tigri*, a term that simultaneously denotes the period in which shellfish may be harvested according to lunar and tidal cycles, as well as the shellfish themselves. We sought to observe practices of subsistence and community resistance in a village marginalized by national development plans and infrastructural projects, as well as by tourism, which thrives along the Souss Massa coast, a region increasingly shaped by environmental changes. Our participatory research developed through a collaborative relationship with local associations, resulting in a range of "visual products" including the mapping of fishing activities, a photography exhibition, and a documentary film, which we had agreed to show at the Tigri Slow Fish festival, one of the most important events of the year in Sidi Bounouar. However, the moment of public restitution during the festival confronted us with several unexpected situations and moments of discomfort. We argue that these moments are analytically significant, as they allow us to reflect on and deepen the methodological discussion of participatory field research.

As anticipated, livelihoods in Sidi Bounouar are challenged by multiple problems. Until a decade ago, before being severely affected by a strong drought, people in the village also practiced subsistence agriculture. In recent years, however, even traditional fishing has become increasingly precarious, undermined by industrial fishing and by the discharge of several desalination plants serving tourism infrastructures across the region. At the same time, new state regulations have begun to challenge long-established foot-fishing practices, and consequently, people in Sidi Bounouar have become more dependent on external markets, which many can scarcely afford. Nevertheless, local identity remains closely tied to the ocean and to *tigri* practices, and the three-day Slow Fish Tigri Festival they have organized represents a way of asserting this connection to the world, despite the apparent loss of interest on the part of Slow Food headquarters (of which Slow Fish is a part) in continuing to support the Wakkad Tigri mussels' *Presidium,* a form of small-scale production officially recognized within the Slow Food network.

The festival was the occasion for us, foreign women researchers collaborating with the festival organizers, to observe our positionalities in the field and to reflect on how previously established relationships were challenged and reshaped during the festival. This was a peculiar situation in many respects: we entered that field as a group, drawing on long-term connections one of us had developed with members of the local community members, yet we were not easily identifiable as a collective actor with a recognizable role. Other guests occupied more legible "expert" positions: as representatives of national Slow Fish branches, as consultants involved in international cooperation projects, or as recognized professors asserting scientific authority. Despite their limited connection to Sidi Bounouar's context, all these groups appeared to hold clear and confident ideas about appropriate "development" solutions or proper Slow Fish practices for the village. These largely revolved around the promotion of *sustainable tourism*, understood as the harmonization of fishing and tourism within a *virtuous blue economy*, grounded in "*good, fair, and genuine food*" (the Slow Food motto) and framed as a driver of *local development*. By contrast, we constituted a heterogeneous group of women researchers, differing in age, academic position, origin, and degree of familiarity with the village, conducting fieldwork on community food practices and environmental change, with the intent to co-produce knowledge with community members about the different drivers of insecurity affecting local livelihoods, their underlying causes, and the bottom-up solutions emerging from within the community, in forms that could be meaningfully communicated to policymakers.

Our presence as researchers, aiming to frame local problems as deeply embedded in broader structural transformations requiring political discussion at regional and national levels, became entangled in dynamics that exceeded our initial intentions.

First, one of us, of Jordanian Palestinian background and wearing a hijab, unexpectedly became a focal figure of the festival. At a time when civilians in Gaza were under continuous and atrocious Israeli attacks, she was welcomed with honours, repeatedly acknowledged in speeches of solidarity, and invited to speak about Palestine. This extraordinary level of attention often left her feeling uncomfortable and overexposed.

Second, the roundtable was designed to showcase the political and cooperation-based multiscalar networks that the local community had built over time. Because we were unable to present our work to the community, the panel lacked contributions grounded in a systematic understanding and mapping of ongoing socio-ecological transformations. Within our team, we discussed whether this marginalisation was related to gender or academic status, positioning us as less credible "solution-makers" than experts affiliated with NGOs and institutions. We also reflected on how the close and long-standing relationships that one of us had established with community members operated in multiple directions: while these ties may have facilitated our sidelining from the formal space of the roundtable, they also enabled informal access to organizers and fostered shared moments of learning through cooking, eating together, and spending time in the kitchen.

A further reflection concerns the village women themselves. The insistence on showcasing their presence in interactions with institutional actors revealed the mainstreaming of gender within international cooperation frameworks. For H., who had organized the roundtable, it was important to prove to both local and international institutions that women were actively involved in what is commonly celebrated as "participation". Yet the women themselves did not share this concern. "It's just blah blah", A. remarked when asked why she had not attended the event. Many of the women H. had expected in the audience had simply forgotten about it. For them, being present in a space they did not perceive as their own was not a priority. They—and we—knew that they had been invited to make up the scene, but not necessarily to contribute with their experience.

These experiences prompted us to reflect on several interrelated issues: gendered spaces and relationality, the language and imaginaries of development, and power relations across scales and moments. During the festival, the small village became a focal point of intersecting interests. At first, our group was asked to ease relations with the Slow Fish network, whose headquarters are based in Italy, and to assist with the writing of project proposals to be submitted to international donors and foreign sponsors. In that context, we found ourselves questioning whether our research roles were deemed relevant or powerful enough, whether our commitment to treating community knowledge as scientifically valid was appreciated yet regarded as carrying less epistemic authority than institutional science, or whether other factors, including gender dynamics, had undermined our presumed advisory role.

Therefore, in this paper, we reflect on the tensions and encounters we experienced in this context, which once again revealed the centrality of relationality in research. These experiences prompted us to question how relationships are practiced and explored in the process of conducting research, co-producing knowledge, and reporting it back to the community. The "micro-village" of Sidi Bounouar offers a situated context for analysing the challenges and possibilities of qualitative, participatory, and visual research, at a time when research practices increasingly call for decolonizing collaborations and for a critical engagement with international cooperation, while also acknowledging and supporting solidarities across spaces. As with all relationships, those built and cultivated with people in the field are shaped by power, which is never static but rather circulates among actors depending on roles, moments, and priorities. Therefore, we seek to unpack power dynamics in the field by exploring how they operate concretely, how they intersect with processes of legitimation, how they may build or hinder reciprocity, and how they can generate misunderstandings or unexpected encounters, turning discomfort into a potentially transformative space. In doing so, we challenge the linear and orderly narratives through which participatory research is often presented and planned. To highlight the transformative potential that discomfort can activate, the following sections will explore various methods (mapping, photography, and video-making), and two moments of research restitution within the festival: the round table discussion and the documentary screening.

We situate these festival moments within interpretive frameworks such as gender perspectives and transcalarity—namely, the relationship between the local context and international actors. By critically engaging with participatory methods, we discuss how discomfort can generate forms of reciprocity that are fundamental to fostering active participation and mutual understanding. In doing so, we reflect on our positionality as inherently intertwined with both the methods employed and our embodied presence in this context.

2 Building and Practicing Participatory Research Between Discomfort and Relationality

It is with critical awareness, often accompanied by the desire to take care of the relationships and people we meet while doing research, that we have engaged in conversations within our team aimed at reflecting on our positionalities and roles within the power dynamics outlined in the previous section. Who carries out research, where, and how, matters (Bilgen et al., 2021). Who we are and how we relate with others and the world in our efforts to build and share knowledge are immersed in, and interact with, power dynamics and hegemonic definitions of the global North and South (Kamal & Courtheyn, 2024). At the same time, practices of academic knowledge production are deeply affected by dominant modes of meaning-making and of building questions, data, and analysis. In affirming the importance of reflexivity and positionality, we are also aware of their problematization and of the limits of how "positionality statements" can often become performative. Gani and Khan (2024) point to how reflexivity in the form of such statements has become a "validity method" that allows researchers to perform "the critical scholar" in ways that remain extremely extractive, as they make local participants "an instrument in the process of self-knowing and indeed *self-asserting*" (Gani & Khan, 2024, 6).

Stating the researcher's positionality and centring it in how research is presented has often become a practice through which the legitimacy of the researcher's authority is reaffirmed once again, placing their experience as a central theme of the research itself. This ends up further marginalizing the contexts and people the researcher has interacted with, using them for the benefit of academic knowledge production by appropriating words, concepts, and ideas extracted from the context and brought into academic spaces to validate research results. To avoid such exploitative and silencing dynamics, we approach the act of stating and reflecting on positionality as a window onto power and relational dynamics that characterize the spaces and places we have inhabited together as researchers during this specific fieldwork.

Research methodologies that explicitly engage with the researcher's positionality in contextual ways are particularly generative. In this direction, these methodologies situate the presence of the researcher within relations and spaces that are shaped by intersected power dynamics. Radical Food Geographies (RFG)—framed as both a theory and praxis—are inspiring in this sense, as they bring attention to power, resistance, and alliances which shape the research process, while acting against the inequities that shape food systems around the world in different ways (Hammelman et al., 2024). RFG, thus, are defined as "a praxis that pushes food systems scholars and activists to pursue work that is simultaneously grounded in theorizing, action and reflection related to the intersections of place and spatiality" (p. 13). Such a perspective interconnects theoretical engagement and action, centring questions of justice both materially and epistemically. Other perspectives that advance similar attentions to intertwined forms and levels of power are included within the wider food studies, where doing things together (Hall et al., 2020) and "following the thing" (Cook et al., 2004) instead of simply observing are intended as "methodological means of tracing the transformations of food; from ingredient, to par-cooked, to creation, to eating" (Hall et al., 2020, 84–85). Furthermore, food studies and geographies have focused on meaning-making through food in original ways, centring food places and practices as key sites where to trace conflicts and frictions.

Non-verbal languages, with their evocative power, are particularly effective in unfolding meanings related to food. In order to access and share food pluriverses (Kothari et al., 2019), convivial and visceral methodologies have been developed, focusing on experiences and emotions in relationship to food (Harcourt, 2025; Sexton et al., 2017), thus acknowledging how food is daily experienced through cultural practices and socio-ecological relationships.

With such approaches in mind, we understand relationality and reflexivity here by "staying with the trouble" of them (Haraway, 2016). In this sense, we have reflected on addressing intellectual and embodied feelings that have emerged in relation to power in the field through feminist lenses that have theorized how starting with the "politics of discomfort" can allow us to work against the reproduction of harmful ways of knowing and ignoring (Chadwick, 2021). Drawing from Chadwick (2021) and Ahmed (2014), we engage with our embodied feelings and epistemic anxieties (Newton, 2022) as emotions produced within personal, collective, and geopolitical relations, and ultimately as spaces for disorientation and reconfiguration of questions and learning. When discomfort is acknowledged as part of the complex processes of meaning-making and knowledge building and sharing, it invites us to think of relationality in research in grounded and rooted terms, as it opens us up to the possibilities of failure and error as epistemic spaces. The implications of such posture are radical, as they demand relating to local communities through participatory methods with an awareness of how economic, political, and cultural power play out in different interactions and spaces.

Within this methodological framework, power dynamics are intended as relational and never static. Adopting a critical perspective to gender norms and roles in this context has proven to be a stimulating entry point for reflecting on how research agendas and cooperation projects are shaped. "Gender" has often become a buzzword used by international funding projects that promote women's empowerment through neoliberal feminist lenses (Zakaria, 2021; Metha, 2002). These frameworks encourage local communities to focus on "gender mainstreaming" issues in order to access funding, making them dependent on international cooperation and aid, while also reinforcing existing inequalities (Wilson, 2015). Such mechanisms impose a "gender agenda" that overlooks contextual meanings of gender, transforming it into an internationally imposed category and as a condition to access neoliberal development projects (Kothari, 2005). In these cases, we can see how "gender" operates both as a criterion and a strategy that local communities navigate and negotiate to secure international funds. By re-elaborating and appropriating narratives of women's empowerment, recipient communities are able to influence donors' priorities while also partially adapting to their languages and conditionalities. In this sense, the performance of gender in international cooperation and research emerges as a key space through which power can be examined in its relational dimension.

Moreover, female researchers often remain confined within gendered dynamics, expected to focus on "women's issues" while broader themes like politics and development are reserved for male colleagues. This segregation persists even in policy-oriented research, where gendered assumptions shape the division of labour, marginalizing feminist perspectives and narrowing the scope of inclusive, comprehensive policymaking (Arora-Jonsson, 2014). These dynamics also raise methodological concerns in participatory research, which requires a significant investment of time and energy, as well as considerable emotional labour to build and sustain meaningful relationships (Foster, 2016).

Within this critical framework, visual languages such as photography and video-making can facilitate relational and trans-scalar interconnection among researchers, food practices, and communities (Varró & Van Gorp, 2021). At the same time, they constitute a site of negotiation and potential harm: visibility can function both as a resource and as a trap, especially for marginalized subjects, when portrayed through exoticizing, objectifying, and patriarchal gaze, and with extractive intents (Brighenti, 2009; Rose, 2001; Sontag, 1997). In order to practice visual methods in a transformative way—both for the researchers and the participants—they must be grounded in a careful approach. Following feminist geography's definitions, care is a process of listening, empathy, responsibility, and reciprocity (Tronto & Fisher, 1990). Listening to participants' discomfort with the camera and its outputs, as well as their curiosity, is essential for building trusting and accountable relationships, which are a prerequisite for relational, participatory, and transformative research. Reporting back, understood as the process of sharing visual outputs, discussing the research findings collectively, and facilitating decision-making with communities involved, represents a crucial phase of decolonial, mutual, and solidarity-based participatory research (Butcher, 2020); yet, this is a long-lasting process that requires an engagement beyond fieldwork per se. As we conceive it, the role of researchers is one of co-construction of knowledge in alliance with specific groups, rather than one of political negotiation with all social actors or with public authorities.

3 Reciprocity as Method, While Navigating Power Dynamics in Researching Foodways

While participatory methods should indeed be approached with a critical lens, they nonetheless offer valuable opportunities for building reciprocity and relationality. Attending to the relationship itself, including moments of misunderstanding, disappointments, or the impossibility of pursuing a planned task, is essential to an ethical and reflective participatory practice. In this section, we present various challenges we faced during the phase of research restitution. As anticipated, our engagement involved multiple participatory methods, including community mapping exercises, preparing food together, taking part in various activities with fishing groups and the women association, and organizing photo and video-documentary presentations on foot-fishing practices. More specifically, we reflect on the power dynamics that shaped the specific moment and space of the Tigri Festival, and on how we responded to emerging tensions while striving to practice reciprocity and care in our engagement with participants. Being in the field together enabled ongoing discussions shaped by our diverse positionalities and relationships with the community and the place. These exchanges revealed the richness and complexity of our collaborative learning process and deepened our connection to the research experience.

3.1 Readjusting Methods While Mapping Environmental Changes Affecting Fishing

We paid attention to forms of discomfort that emerged as we planned and conducted the research. Our intention was to co-create a process of "mapping meanings" of space and movement (Hohenthal et al., 2017) to understand the broader environmental, socio-ecological and structural transformations that had shaped the lives of inhabitants of the Souss Massa region. However, as with many forms of action-research that engage with struggles not directly experienced by the researchers themselves, this proved to be a challenging exercise. While community mapping is a creative process that enables multivocality and the emergence of counter-narratives, the specific ways of engaging with the community and eliciting stories about relationships with the sea and coastline unfolded through trials, mutual understandings, and the gradual building of trust. When we first organized a focus group to discuss fishers' experiences and the local economy's transformations, we invited participants to draw timelines of their fishing stories. This moment confronted us with the inequalities and power dynamics embedded in academic research methodologies and with our own positionalities within the relations enacted by research. Many fishers felt uncomfortable using pen and paper and were uncertain about what was expected of them. Surely, our inability to speak Tamazight hindered our full understanding of their conversations about fishing practices in this part of the Atlantic coast. With the help of younger community members, however, we were able to bridge some of these linguistic and knowledge gaps. Intergenerational dynamics played a key role in this process: elders shared their stories, while younger mediators translated and illustrated them, enabling a collective practice of storytelling and mapping.

The mapping exercise revealed three key themes: (1) the accumulation of sand along the coast near Sidi Bounouar, which, as one fisherman noted, "kills the sea" by eliminating shellfish habitats and reducing marine biodiversity; (2) the decline in fish stocks, due both to climate change, revealed by the rising seawater temperature, and to industrial fishing; and (3) the decreased recognition and economic viability of foot-fishers. Regarding the first issue, a shared hypothesis emerged: the coastline from Agadir to Sidi Bounouar, once shaped by an ocean current that carried sand away from the shore, is now impacted by the expansion of the port of Agadir located approximately 100 km up to the north. According to the fishermen, this expansion may have disrupted that current, leading to sediment accumulation and gradual burial of rocky habitats. As fishers recalled their mobilities over the years, they explained the progressive erosion of their livelihoods in relation to these changing marine and coastal environments, where fish stocks are in continuous decline. In their view, local authorities are aware of these issues but uphold the strategic importance of the Agadir harbour. Similarly, while recognizing the ecological and economic sustainability of foot and artisanal fishing, authorities continue to prioritize industrial fishing.

As a matter of fact, foot-fishing is absent from the official redefinition of the fishing economy. Regulations governing different forms of fishing treat foot-fishing inconsistently: sometimes as legitimate work, other times as leisure. As a result, foot fishers lack formal representation and public support, as authorities officially recognize only artisanal boat fishing and industrial fishing. In the absence of clear regulations, foot-fishers face increasing difficulties in legally selling their catch. During various informal contexts, rather than in formal meetings, and particularly when one of us accompanied them to some administrative offices, fishers described numerous bureaucratic obstacles. The challenges they face are shaped by environmental pressures, often of anthropogenic nature, intertwined with bureaucratic constraints that further undermine the livelihoods of the poorest fishers.

The maps (Fig. 1) thus reflect changes in fishermen's lives as documented through their life stories. They were made possible by the trust extended to us, which granted access to narratives of marginalization and structural disempowerment. The maps bring to light issues that had not previously been addressed in academic studies or technical reports. For this reason, our inability to bring public attention to them during the festival, in front of public authorities and even local association leaders, was especially disheartening. Nonetheless, the relationality and responsibility established through this research do not end with the conclusion of the research funding. Because the Tigri Festival represents a rare moment in which public attention converges around community fishing and seafood, we intend to reintroduce this study and more effectively bring it into the public arena at future festivals.

3.2 Unsettling Participatory Research. Reimagining Methods Towards Conviviality

Participatory research offers the possibility of subverting power dynamics and creating unexpected opportunities. Initially, we had planned a photovoice process with the village women, but they asked us instead to organize a meeting to exchange ideas for new projects and help them write them. Over time, we recognized that cooking provided the most pleasant medium for exchanging ideas in conviviality. A few days before the festival, we therefore organized a cooking session to share reflections and practices related to food system transformations and their aspirations for the new projects. The event quickly became one of the large, periodic gatherings that women host in the mosque house. We made clear that we would cover the cost of the food, as the meeting was our initiative. When purchasing ingredients together, the cooperative women suggested

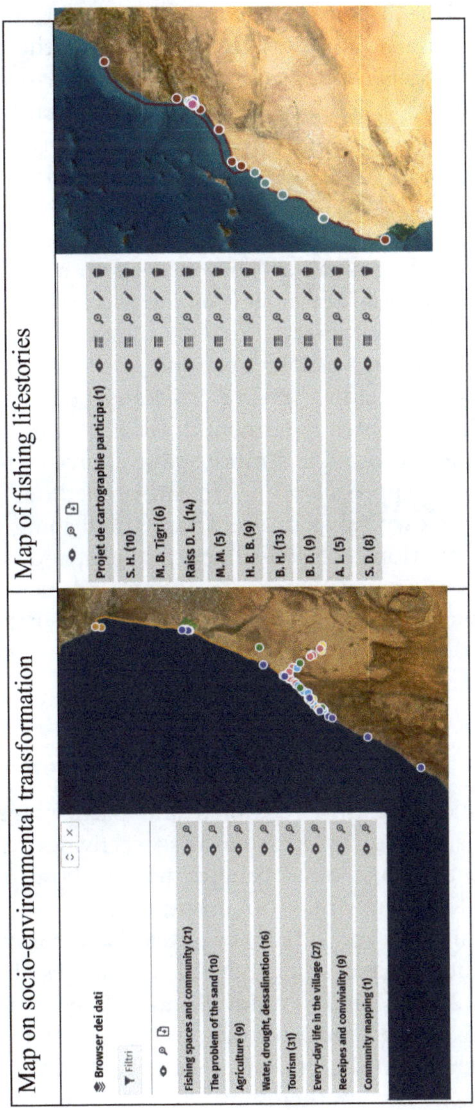

Fig. 1 Two maps presented at the roundtable (Beatrice Ferlaino, 2024)

preparing *baddaz* (corn couscous) without meat to save money. Eventually, about eighty people shared the meal. What began as a workshop for cooperative members became an occasion to provide food to many women and children, including attendees arriving for the forthcoming festival, with some portions taken home for other family members.

Although this moment differed from our initial expectations, it proved invaluable for building reciprocity. The community appreciated that we invested time, labour, and resources in the shared experience. After the meal, despite fatigue, the cooperative women remained to speak with us. They were clear: the sea no longer provides enough shellfish, making mussel-based projects unsustainable. They wished to acquire new skills and to challenge the essentialized image of themselves as "mussel pickers", a depiction widely legitimized through international cooperation projects. While they do not wish to abandon shellfish gathering altogether, they asked us to develop a sewing project, producing kaftans and local *addals* (traditional women's garments), or a cooking and catering initiative, both intended to provide a more stable source of income.

Through a reciprocal approach to participatory methods, the women of Sidi Bounouar took ownership of the process, reshaping it according to their needs and aspirations. They assumed the lead, and we welcomed their initiative, recognizing that they had a deeper understanding of how to cultivate conviviality in this context. This experience enabled an open and sincere sharing of ideas about changes in the food system and their perspectives on the future.

3.3 Relational Gazes: Building Trust Through Visual Methods

Conducting fieldwork through a relational and participatory approach entails navigating languages, places, and subjectivities, while being attuned both to practices that build bridges and to those that generate fractures and discomfort. The visual tool prompts crucial questions about the meaning of visibility, the accessibility of physical and emotional spaces, the colonial and orientalist gaze, and gendered experiences of being represented (Tolia-Kelly, 2012). In fact, Sidi Bounuar's people, depending on their positionality and personal histories in a context of unbalanced power relations, experienced the presence of the camera very differently. In general terms, men offered themselves for shots or invited to use the camera, seeing photography as a proof of the value placed on their actions, words, lives, and roles. On the contrary, for many women the camera symbolized intrusion and a lack of respect for personal boundaries, often generating discomfort, as described in the opening vignette. This discomfort, expressed by participants and mirrored by the epistemic anxieties and unease experienced also by those operating a camera on the field, confirmed how, in Sidi Bounouar "looking" ("seeing" and "being seen") is not only a culturally variable activity, one that is subject to ethnographic inquiry, but also one shaped by processes of socialization, including gender norms and relationships with technologies and images (Chalfen, 2020).

Building trust is essential to defining the boundaries of visibility and to reframing photography and video-making as acts of mutual exchange rather than extraction. This ethical and theoretical stance takes shape through a gradual relationship with places, a negotiated proximity between the camera and its subject, and the interplay between the photographer's gaze and the visions of the community involved. This process is reflected in the visual archive of the project: early photos were wide, panoramic, suggesting distance and caution. Over time, through a process of mutual (often non-verbal) understanding and empathy, pictures became more intimate, zoomed-in, and warm—taken *with* places and people (Hunt, 2014; Lorimer, 2010). A symbolic moment in this direction was S.'s gesture: while she avoided cameras and journalists during the festival, she later approached us, proudly dressed in her traditional Amazigh attire, asking to be photographed and saying, "This one is for you!". Nevertheless, most of the photographs do not depict faces, but rather hands: hands at work, cleaning fish, cutting octopus and vegetables, lighting fires, and cooking.

Sharing images and video throughout the process invited participants into the researcher's perspective, opening up spaces for dialogue and negotiation. As the use of the camera was made explicit and accessible, decisions about what to photograph or film were frequently shared: participants often guided our gaze, suggesting what should be captured. The visual material thus came to constitute a collective archive of gazes—those of the researchers and the community. The challenge of working with such an archive lies in building a narrative that does not claim neutrality but unfolds the positionality of the visual researcher within our team and the collaborative relationship established with local participants. For this reason, during the festival, we decided to present a photographic exhibition and a short film[1] as forms of responsible narration and as careful modes of restitution.

Attending a screening about one's land and community involves a complex interplay of ownership over the project, as well as the risk of estrangement and discomfort in relation to the researcher's gaze. Thus, the editing phase must reflect a situated awareness of both the audience and the context of the screening: in our case, a multilingual Slow Fish festival, with an audience largely unable to read. We therefore sought to give back a narrative attentive to the ecological, economic, and social pressures shaping the territory and its food practices, while also rendering visible the intimate dimension of cooking, eating, and dancing together, and the different spaces and gender roles involved (Fig. 2). We ultimately set up the photographic exhibition in the main hall of the festival, where meals were also served (we later learned that the exhibition had been made permanent by the association), and screened our video on the first evening. The reactions elicited by these two forms of media were as powerful as they were unexpected. The organizers helped selecting the photographs for the exhibition (Fig. 2c.) and were the first to view a preview of the documentary preview, responding with both surprise and emotion. During the festival, people engaged with curiosity, and community members, including women, engaged warmly and emotionally, at times laughing together at scenes they found amusing. The sequences displayed everyday life moments (storytelling, cooking, walking), where our local partners were the main characters and guides. "This is us, this is our life", said one of the local women at the end of the screening. This moment of careful and genuine sharing made our presence more intelligible, while also casting new light on community practices and the territory from an external perspective that was nonetheless considered meaningful and respectful. Through this emphatic negotiation, we were able to give back to the community what made the most sense out of our visual work, that is, a new, mobile, multimodal space where their stories are told.

[1] The full version of the short film is available at the following link: https://www.geovisum.com/2026/04/15/il-mare-prende-il-mare-da/.

Fig. 2 Photos documenting the ecological and social dimensions of foodways in Sidi Bounouar (Ginevra Montefusco, 2024) (a) Elder fisherman picking mussels during Tigri. Talat Bouâaza, (b) Artisanal fishing port of Sidi Boughdail under demolition, (c) Selecting pictures for the exhibition in Sidi Bounouar, (d) Eating mussels couscous at the *Maison de l'homme blue* (I.'s house)

3.4 *Performing Internationalization Between Research Legitimation and Cooperation Expectations*

Another theme worth reflecting on concerns our role in the "internationalization"of the village, and the type of roles and functions that we could meaningfully support in that process. The village identity, within the festival, is closely tied to its status as a *Presidium* of the Slow Fish/ Slow Food network. For us, being present in that context meant navigating a dual form of governance and recognizing how certain practices and discourses we observed were shaped by the need to align with both frameworks.

It was, for instance, interesting to observe the official narratives constructed by Slow Food around Sidi Bounouar and the activity at the centre of its focus: the harvesting of *Wakkad* mussels by women fishers. These narratives portray a community of women foot-fishers who harvest shellfish and maintain a profound, daily, lifelong, sustainable relationship with the sea, striving to keep alive the tradition of hand-fishing. The *Presidium* product must also adhere to a specific smoking process, using stone ovens, particular types of wood, and selected herbs, as strictly defined in documented protocols.

Although lived realities are far less romantic and more complex—for instance, hand-fishing is practiced by both women and men, and the smoking process does not strictly adhere to the protocols (as was noted during the official demonstration at the festival, starting with the fact that the mussels were not wild, given their increasing scarcity, but purchased elsewhere)—the local population still embraces the performance of this narrative. Belonging to a global network generates specific local power dynamics: it opens opportunities, fosters relationships, and can legitimize the otherwise unheard voices of foot-fishers in their dealings with local authorities.

This relational process of legitimation led the community of Sidi Bounouar to adopt categories promoted by the international cooperation—such as "gender mainstreaming" or "sustainability" discourses—as ways of presenting themselves. Yet, at the same time, we too played a role in this internationalization process, which functioned as a form of mutual legitimation. After all, were we not granted legitimacy precisely by virtue of our status as foreigners, embodying the festival's international audience? And were not our local interlocutors publicly empowered through their association with an international research project? This is why we were often invited to offer brief greetings, present awards, sit in the audience's front row, and be visible in front of regional television cameras, thereby performing the internationality of the Tigri festival. In all these moments we felt that we were positioned in visible roles not because of any specific contribution we were expected to offer, but because our presence itself carried legitimating weight: at the risk, as the vignette suggests, of remaining a visible but silent audience.

We found ourselves caught between opposing dynamics. On the one hand, we embodied a form of internationalization that contributed to legitimizing the festival and, by extension, the organizing group. On the other hand, we did not perceive ourselves as part of the broader international group whose interventions were criticized by festival participants during the roundtable, precisely because of their detachment from the local context. Despite their stated intentions, these international and institutional actors largely promoted development in primarily economic terms, albeit framed as sustainable. Their approaches rested on the assumption that tourism could add value to local practices—such as Wakkad mussel fishing—by attracting visitors seeking direct, on-site experiences of production, thereby encouraging the community to maintain these practices in response to tourist-driven demand.

However, such proposals overlooked key issues that we had come to understand through our fieldwork. These included the environmental changes undermining foot-fishing practices, the impacts of tourism infrastructure development, and the lack of licences allowing foot-fishers to legally sell their catches. Moreover, the implementation of community-based tourism would entail significant challenges, requiring substantial financial resources and entrepreneurial skills to host visitors and expand food provision, while also contending with inadequate water and sanitation infrastructure. The extension of *tigri* practices throughout the year, despite their seasonal nature, further risked placing strain on the community's spatial organization and generating new vulnerabilities rather than empowerment.

These concerns were explicitly raised by local participants during the discussion that followed the roundtable speeches. Participants also criticized the organizers for failing, in that specific moment, to provide space for community representatives to articulate their perspectives on the environmental and infrastructural changes affecting fishing practices, marine life, and local livelihoods. By drawing attention to the vital interdependence between water-based (fishing) and land-based (agro-pastoral) practices, those who were able to intervene offered a grounded understanding of foot-fishing as embedded within broader food ecologies and their ongoing transformations.

As a result, we occupied an in-between position—distinct from that of other foreign participants, yet one that was likely not fully legible to many of those present. Indeed, we aligned with local contributions that shifted the discussion away from development narratives centred on tourism investments and back toward everyday lived experiences and material challenges.

3.5 *The Festival: A Shared Time Across Different Positionalities*

As a research team, being together in the context of Sidi Bounouar was characterized by the plurality of our positionalities. Our collective presence in the village during the festival was not homogeneous but rather grounded in multiple temporalities and relationships with the local community, the wider Moroccan context and the region at large. While one of us had developed a long-term research presence in the field, another had long-standing knowledge of other regions in Morocco, and others were encountering both the community and the country for the very first time.

In general, we did not experience difficulties in getting to know community members, largely due to their strong culture of hospitality and to the many shared activities we engaged in together, often collaborating without the need for verbal communication. Nevertheless, linguistic challenges and cultural differences were present; issues of language and embodiment proved crucial to our ability to communicate, to understand social interactions and cultural norms, and to be understood by the local community. Moreover, as a research team composed of members with diverse cultural and experiential backgrounds, we frequently noticed and discussed the different kinds of questions we had, and the responses, and reactions we encountered, comparing our situated understandings and interpretive perspectives.

While most of us spoke in French, one of us mainly communicated in Arabic. In a context where people mostly spoke Tamazight, Arabic was not immediately accessible to everyone and in all interlocutions. It did represent, however, an added value and an access point to meaningful conversations with those community members who spoke *fusha* (formal) Arabic. In fact, in specific moments when French became difficult to use or people did not speak it, having the medium of Arabic as well allowed connection and interaction between us and the people we encountered.

As already mentioned, the fact that the Arabic-speaking member of our team was primarily presented by the festival organizers as Palestinian constituted another element that elicited expressions of solidarity and closeness from participants. At the same time, the emphasis placed on this aspect (through repeated invitations for her to address the audience) offered us an opportunity to reflect together on how local organizers, village inhabitants, and festival participants found ways to make Palestine present among them. Inviting our Palestinian research team member to cook a Palestinian dish on the final day of the festival, asking her to speak publicly, or filming her presence, were not acts grounded on her individual relevance or role within the research, but rather on what she came to represent in that specific moment: a shared political awareness and a form of collective solidarity. The diversity of our research team thus became a significant element shaping our relationships and interactions within the festival space, as different moments and encounters were structured around the team member most actively engaging with the community at a given time.

This dynamic enabled us to develop and practice a more nuanced and complex presence in the field, as well as a heightened awareness of the relationships we were part of, and of our own positionality. Meeting other international researchers and practitioners during the festival further highlighted the power dynamics embedded in the performance of such relationships, whereas "internationals" were consistently seated at separate tables in restaurants and in front rows during cultural events: a form of hospitality that, rather than producing comfort, generated a sense of unease in our team.

Engaging with those emotions, however, became an analytical space through which we could recognize how such representations and performances were also arenas in which different actors enacted and negotiated their agency toward specific goals and relationalities. It was precisely within this discomfort that we became more aware of the legitimation processes produced through our involvement and centrality in particular spaces.

4 Conclusions

This paper has focused on a specific moment of fieldwork conducted by a team of women researchers from an Italian university in the Moroccan coastal village of Sidi Bounouar. The fieldwork formed part of a broader research project on the foodways of communities exposed to multiple, intersecting pressures—socio-ecological in nature and operating across different scales—that increasingly undermine their livelihoods.

The phase examined here coincided with the Tigri Slow Fish festival, an international event that is nonetheless deeply rooted in local traditions, to which the researchers contributed and within which they were expected to present a form of restitution of their participatory and visual research. Not all actions unfolded as planned: the course of events, the competition among different performativities in search of visibility, the international political context, and other factors required, on the one hand, multiple exercises of adaptation on our part, and on the other, sustained discussion and methodological reconfigurations, which we have sought to share in this paper.

Many questions emerged during and after the fieldwork, raising issues of legitimation, including the following: What do we mean by *participation* in research? Who is expected to participate, and in whose projects? Who are we to propose research projects in contexts that are not our own, and to ask local communities to participate in them? Instead, as external agents, how can we better participate in and potentially serve proposals informed by groups that are often marginalized? And how can we effectively support local causes while avoiding manipulation and opportunism by local or external leaders that may undermine the originally shared objectives?

What emerged from the individual and collective moments of research described above speaks, on the one hand, to critiques of participatory methods, and on the other, to the need to identify ways of genuinely and consistently practicing relationality and participation in convivial moments, while engaging with interlocutors to learn about their lives, territorialities, environmental changes, food cultures, ancestral knowledges and regional planning, in order to develop analytical readings of causality from the community's perspectives.

Considering discomfort as a generative space when practicing participatory methods essentially means being connected both with others and with oneself. Relationality is fundamental in this process. The co-construction of methods, questions, and perspectives begins with mutual understanding: experiencing discomfort when proposing practices (such as drawing or speaking), when reflecting on one's posture, or when recognizing one's own limits (such as language and body language) proved to be a valuable starting point for questioning how to adapt our research practices, make ourselves intelligible to the people we were working with, and develop creative spaces for communication and interaction.

Conviviality and relationality thus have profound implications that extend beyond the form of the spaces shared during research encounters. They concern the meanings and knowledges that are produced when relationships are taken seriously. Our encounters with the people of Sidi Bounuar, during the festival and beforehand, called on us to practise continual reflexivity, self-critique and self-awareness, and to remain mindful that our presence and positionalities, as we move across global and local spaces, are always embedded in power dynamics.

The post-development lens adopted in our research project has allowed us to critically interrogate these dimensions and analyse how they speak to global neoliberalism, climate change, regional politics and food crises. Food has carried, in this context, multiple processes: representations enacted by different actors, imaginations of different futures, and the articulation of policies and discourses concerning the present and past of the village and the wider Sous Massa region. Reflexivity on positionality has therefore been approached here as an entry point for articulating multiscalar power dynamics within spaces and interactions involving international cooperation, academic research, local communities, and institutions.

Sharing food—through recipes, cooking and eating—was significant throughout different moments of the research project. Cooking and eating together was the perfect setting for informal storytelling and meaningful conversations with the women about their needs, challenges, aspirations, and projects. We came to understand that the structured activities and forms of discussion and meeting we had initially planned would not have fostered real complicity or interest on the women's side, nor would they have offered a space of comfort and intimacy in which they could truly share their stories and ideas. Cooking and feasting together, in resonance with decolonial epistemic practices, enabled conversations and exchanging during and beyond the shared moments of gathering, and proved to be a stimulating and inclusive space and experience for the women with whom we sought to engage.

This closeness enabled a grounded understanding of socio-environmental transformations, connected both to historical changes along the shoreline and to the economic transformations of the Souss Massa region. Participatory mapping carried out through life stories and by walking together along the shore while they were fishing, observing and discussing everyday challenges, made visible the experiences of this population, which are often excluded from dominant narratives about the region, yet are crucial for grasping its structural and ecological transformations. Finally, the sharing of photos and videos during the festival not only fostered a clearer understanding of the role of research among community members but also created a mirror in which they felt recognized and valued.

References

Ahmed, S. (2014). *The Cultural Politics of Emotion*. Edinburgh University Press.

Arora-Jonsson, S. (2014). Forty Years of Gender Research and Environmental Policy: Where Do We Stand? *Women's Studies International Forum*, 47/B, 295–308.

Bilgen, A., Nasir, A., & Schöneberg, J. (2021). Why Positionalities Matter: Reflections on Power, Hierarchy, and Knowledges in "Development" Research. *Canadian Journal of Development Studies/revue Canadienne D'études Du Développement*, *42*(4), 519–536. https://doi.org/10.1080/02255189.2021.1871593

Brighenti, A. M. (2009). *Territori migranti. Spazio e controllo della mobilità globale*. Ombre Corte.

Butcher, S. (2020). Research Solidarity? Navigating Feminist Ethics in Participatory Action-Research in Kathmandu, Nepal. *Gender Place & Culture, 28*(4), 497–518.

Chadwick, R. (2021). On the Politics of Discomfort. *Feminist Theory, 22*(4), 556–574. https://doi.org/10.1177/1464700120987379(Originalworkpublished2021)

Chalfen, R. (2020). Looking Two Ways: Mapping the Social Scientific Study of Visual Culture. In *The SAGE Handbook of Visual Research Methods* (pp. 37–56). SAGE Publications.

Cook, I. (2004). Follow the Thing: Papaya. *Antipode. A Radical Journal of Geography, 36*(4), 642–664. https://doi.org/10.1111/j.1467-8330.2004.00441.x

Foster, V. (2016). *Collaborative Arts-Based Research for Social Justice*. Routledge.

Gani, J. K. & Khan, R. M. (2024). Positionality Statements as a Function of Coloniality: Interrogating Reflexive Methodologies. *International Studies Quarterly, 68*(2), sqae038, https://doi.org/10.1093/isq/sqae038

Hall, S. M., Pottinger, L., Blake, M., Mills, S., Reynolds, C., & Wrieden, W. (2020). Food for Thought? Material Methods for Exploring Food and Cooking. In *Mundane Methods*. Manchester University Press. https://doi.org/10.7765/9781526152732.00012

Hammelman, C., Levkoe, C. Z., & Reynolds, K. (Eds.). (2024). *Radical Food Geographies: Power, Knowledge and Resistance* (1st ed.). Bristol University Press. https://doi.org/10.2307/jj.9692676

Haraway, D. (2016). *Staying with the Trouble: Making kin in the Chthulucene*. Duke University Press (Experimental futures. Technological lives, scientific arts, anthropological voices).

Harcourt, W. (2025). *Conundrums of Care: Feminist Entanglements in Critical Development Studies*. Bloomsbury.

Hunt, M. A. (2014). Urban Photography/Cultural Geography: Spaces, Objects Events. *Geography Compass, 8*(3), 151–168. https://doi.org/10.1111/gec3.12120

Hohenthal, J. M., Minoia, P., & Pellikka, P. K. E. (2017). Mapping Meaning: Critical Cartographies for Participatory Water Management in Taita Hills Kenya. *Professional Geographer, 69*(3), 383–395.

Kamal, A., & Coutheyn, C. (2024). Where is the North? Southern ruptures in decolonizing theory. *Political Geography*, 108. https://doi.org/10.1016/j.polgeo.2023.103029

Kothari, U. (2005). Authority and Expertise: The Professionalisation of International Development and the Ordering of Dissent. *Antipode. A Radical Journal of Geography, 37*(3), 425–446. 5, https://doi.org/10.1111/j.0066-4812.2005.00505.x

Kothari, A., Salleh, A., Escobar, A., Demaria, F., & Acosta, A. (2019). *Pluriverse: A Post-Development Dictionary*. Tulika Books.

Lorimer, J. (2010). Moving Image Methodologies for more-than-Human Geographies. *Cultural Geographies, 17*(2), 237–258. https://doi.org/10.1177/1474474010363853

Newton, L. (2022). Epistemic Anxiety and Epistemic Risk. *Synthese, 200*, 324. https://doi.org/10.1007/s11229-022-03788-7

Rose, G. (2001). *Visual Methodologies*. Sage Publications.

Sanders, R. (2007). Developing Geographers Through Photography: Enlarging Concepts. *Journal of Geography in Higher Education, 31*(1), 181–195.

Schwartz, J. M., & Ryan, J. R. (2021). *Picturing Place: Photography and the Geographical Imagination* (1st ed.). Routledge. https://doi.org/10.4324/9781003268260

Sexton, A. E., Hayes-Conroy, A., Sweet, E. L., Miele, M., & Ash, J. (2017). Better than Text? Critical Reflections on the Practices of Visceral Methodologies in Human Geography. *Geoforum, 82*, 200–201. https://doi.org/10.1016/j.geoforum.2017.03.014

Sontag, S. (1977). *On Photography*. Penguin Books.

Tolia-Kelly, D. P. (2012). The Geographies of Cultural Geography II: Visual Culture. *Progress in Human Geography, 36*(1), 135–142.

Tronto, J. C., & Fisher, B. (1990). Toward a Feminist Theory of Caring. In E. Abel & M. Nelson (Eds.), *Circles of Care* (pp. 36–54). SUNY Press.

Varró, K., & Van Gorp, B. (2021). Fostering a Relational Sense of Place Through Video Documentary Assignments. *Journal of Geography in Higher Education, 45*(1), 63–86. https://doi.org/10.1080/03098265.2020.1803223

Wilson, K. (2015). Towards a Radical Re-appropriation: Gender, Development and Neoliberal Feminism. *Development and Change, 46*(4), 803–832. https://doi.org/10.1111/dech.12176

Zakaria, R. (2021). *Against White Feminism: Notes on Disruption* (1st ed). W. W. Norton & Company, Incorporated.

Index

P. Minoia (ed.), *Foodways As Sites of Knowledge and Transformation*, Emplacing Food, https://doi.org/10.1007/978-3-032-28126-5

Kolibri 360 GmbH
Ferdinand-Jühlke-Straße 7
99095 Erfurt, Deutschland
produktsicherheit@kolibri360.de